12.95

How John Deere Tractors & Implements Work

Roy Harrington

Published by the
American Society of Agricultural Engineers
2950 Niles Road, St. Joseph, Michigan

Credits:
USDA, Natural Resources Conservation Service: Photo on page 5a.
Deere & Company: All other photos.

GreenStar, Hi-Cycle, Hydra-Push, John Deere, MaxEmerge, PowrQuad, and VacuMeter
are trademarks of Deere & Company.

How John Deere Tractors & Implements Work
Editor: Richard Balzer
Project Manager: Melissa Carpenter Miller

Library of Congress Cataloging-in Publication Data
Harrington, Roy
How John Deere Tractors & Implements Work

Summary: How implements and tractors work to produce crops and livestock is described
and illustrated.
1. Agricultural machinery — Juvenile literature. 2. John Deere tractors — Juvenile literature. 3. Farm Equipment — Juvenile literature. 4. Farms — Juvenile literature.
[1. Agricultural machinery. 2. John Deere tractors. 3. Farm Equipment.] I. Title
1997   631.3

ISBN 0-929355-88-1
LCCN 97-75094

Printed in the United States of America

Table of Contents

Why Are Plows Used Less Now?

A moldboard plow turns over black topsoil to cover residue from a recently combined corn field.

John Deere got his start by developing a self-scouring steel plow. It made it practical for farmers moving west to till the sticky soil and grow crops for food where there had only been prairie grass. Farmers till or stir the soil for two main reasons. They first prepare the soil so the seeds they plant will sprout and grow well. Secondly, weeds need to be kept out of the field so they do not compete with the crop for water, sunlight, and minerals.

Fifty years ago, farmers plowed much of their land in the fall after harvest so they would have less to do in the spring. Wind erosion of the exposed topsoil was a problem in some areas. Erosion from rain was a problem on land with hills or long slopes. Some erosion was considered a trade-off for getting the maximum yields from the next crop.

Nature takes a long time to build topsoil. In many areas, erosion was taking away the topsoil faster than it was being replaced. Farmers, who wanted to till in the fall, were encouraged to leave much of the residue or remains of the previous crop on top of the tilled soil. Residue

Spring trips are available on large plows for protection when a hidden rock or stump is hit. The plow bottom trips, rises over the obstruction, and returns to work without ever stopping the tractor.

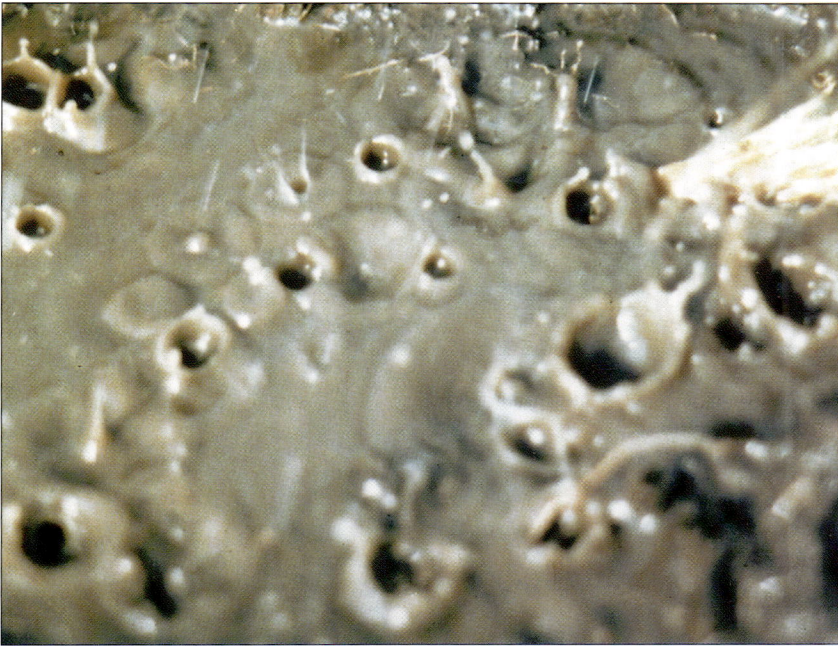

These raindrops are hitting soil that is not protected by a cover of residue The soil particles loosened by the splashing rain soon form muddy water that flows down the slope causing erosion.

by field cultivation and planting. This method will leave about 30% of the soil covered by stalks, roots, and leaves after planting.

Conventional tillage, such as moldboard plowing, leaves less than 15% of the soil covered by residue at planting time. Since 1989, more than half of all cropland has used some form of reduced or mulch tillage that leaves more than 15% soil coverage at planting.

The practice of no-till has grown from 5% of cropland in 1989 to 15% in 1996. No-till means no tillage implement is used prior to planting. Instead, chemicals are used just prior to planting to kill any weeds. The planter or drill does till the soil when its disks open a furrow to place the seed at the desired depth.

Farmers are some of our best environmentalists. Each season they must decide what technology will both grow the best crops and sustain the land so their children and grandchildren can also grow food.

acts like a snow fence to stop wind erosion. Residue reduces water erosion by shielding the soil from raindrops that splash and loosen the soil. It also makes many small dams to slow down the flow of runoff water.

Many farmers started using chisel plows instead of moldboard plows in the fall so residue from the previous crop of corn, sorghum, or wheat was left to protect the soil. In the spring they may use shallow disking to cut the stalks, followed

Soil properly tilled in the fall leaves much of the residue in place and opens furrows for the rain to enter the ground.

On average, each farmer in the United States grows enough food to feed 101 people in our country. In addition, enough wheat and other foods are exported to feed 28 people in other countries.

What Digs and Loosens the Soil?

Chisel plows are the most widely used tillage tool in the fall. They are also used in summer fallow areas where wheat is grown.

Plants germinate and grow best in soil that is not too hard and dense. It can become compacted if tractors and combines are used when the ground is too wet. Soil has a structure something like several tasty foods. Air spaces are needed in the soil, like those seen in a slice of bread, to hold water. The ideal soil is loose and crumbly like a fresh blueberry muffin. Frequently soil is more like a brownie that dried out before it was eaten. Some soil can become as tough and dense as a stale bagel.

The chisel plow has replaced the moldboard plow as the preferred tillage tool in the fall after combining corn or sorghum. It digs up to a foot deep to loosen the soil. This helps plants grow long roots to reach moisture and minerals. It leaves furrows for water to gather and enter the soil for future use. It leaves most of the previous crop's residue on the surface to protect it from wind and water erosion.

V-Rippers work much like chisel plows but may dig almost 2 feet deep. Rippers are most effective when used in the fall in soil that is dry enough to shatter. Since rippers dig much deeper than chisel plows, they take about twice as much power for similar rates of work.

A V-Ripper is the preferred fall tillage tool when the soil is compacted rather deep.

The double-point shovel, shown on the left, is available on both chisel plows and field cultivators. Sweeps, shown on the right, are another popular choice. The spring trip, shown above, is available on both tools for protection from rocks and stumps.

Chisel plows have more than 2 feet of clearance under the frame to pass over heavy residue without clogging. V-Rippers dig deeper so they have even greater clearance to prevent plugging.

Tillage tools used to prepare seedbeds have points spaced closer together and dig less deep. Field cultivators are the third, and most popular, member in the family of tillage tools that dig. The final tool is usually a coil-tine or spike-tooth harrow attached to the field cultivator. The harrow levels the soil and residue for planting.

The field cultivator and harrow also help reduce competition from weeds. They dig out most weeds that have started. They also help mix in chemicals used to prevent new weeds from coming up.

Double-point shovels dig deep and provide furrows to capture rainfall. Sweeps do not dig as deep but provide a

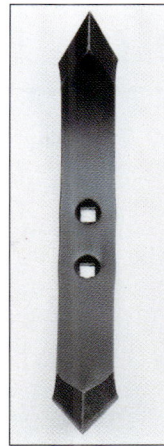

wider strip that is tilled for better weed control. Either choice provides maximum retention of residue cover.

Sweeps are also used in summer fallow to control weeds while retaining soil moisture. The land is left idle one season to collect moisture to grow a better wheat crop the next year.

Twisted shovels are available for conditions where more covering of residue is desired. They act something like small moldboard plows.

Field cultivators with harrow attachments provide a good seedbed whether in reduced tillage or in clean tillage.

If you want to go fishing, look in fresh tilled soil for earthworms. The amount of earthworms you find helps tell how healthy the soil is. Worms aerate and fertilize the soil so plants grow better.

What Rolls and Slices Trash?

Disks are the best tool to use when trash or residue needs to be sliced to reduce plugging in future tillage or planting operations. As shown here, shallow disking leaves much of the residue on the surface.

Disks have long been a favorite tool for tillage due to their versatility. They can be operated deep as the first tillage operation or shallow for seedbed preparation. Many design options are available depending on the soils and crops in which they are used. Various disk widths and weights are available to match tractor sizes.

A spacing of 9 inches between blades provides a very adaptable disk for various conditions. A narrower spacing is suitable for shallow operation. A wider spacing works best for deep tillage. Disk blades are circular but some have cutout edges to keep trash from being pushed ahead of the blades instead of being cut.

Scrapers are used on the inside of the blades to remove sticky soil from the blade. Disks with thicker blades and spring-cushioned disk gangs are available for operation in rocks.

Several tillage and planting implements use rolling coulters to cut residue so it does not wrap around the tool. Most plows have coulters and many rippers use coulters to slice trash. Coulters are also used on many planters and no-till drills to reduce plugging.

Most coulters are simply flat disks with edges sharpened to cut trash and make a narrow furrow. Fluted or rippled coulters till a wider strip. Coulters can also be set at an angle to clear residue from a wider path.

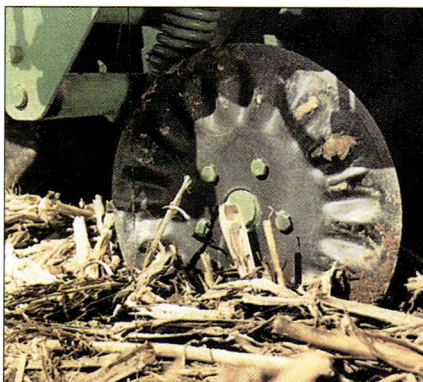

This ripple coulter is mounted to the main frame of a planter to slice corn stalks. It also tills a 1-inch wide strip in front of planter openers.

A Mulch Finisher provides an excellent seedbed in a single pass in soybean ground.

Several tillage tools have been developed to reduce soil erosion by leaving more of the residue on top. These tillage tools reduce the number of trips across the field by combining a disk at the front with digging tools. Disk Rippers, used in the fall, can till more than a foot deep to break up compacted soils.

Mulch Tillers use straight coulter blades in front, followed by chisel plow shovels or sweeps. Most of the residue is retained but sliced to avoid clogging in future operations.

Mulch Finishers use a disk gang in front, followed by wide field cultivator sweeps to get complete width tillage. Mechanical weed control is excellent. Future weed growth can be controlled by adding nozzles to spray chemicals.

Single-disk or double-disk openers are used on most drills for planting wheat, oats, rice, and barley.

Double-disk openers with adjacent gauge wheels have been used on planters for more than 20 years. Disk openers for drills and planters slice and retain the additional residue needed to reduce soil erosion. They also open a furrow just wide enough to deposit the seed while maintaining soil moisture so the seed will germinate.

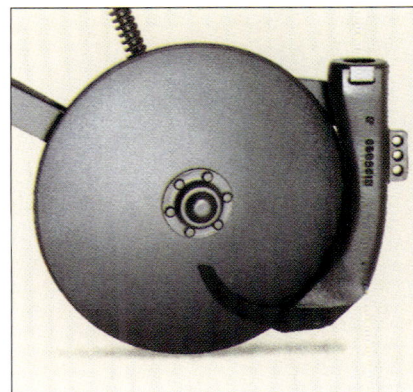

The basic design of the single-disk opener remains similar to that used with horse-drawn drills. It cuts trash and penetrates hard soils.

Nature compensates for the erosion resulting from farming. Each year enough new topsoil is developed in a 40-acre field to fill about 13 dump trucks.

How Are Corn and Cotton Planted?

A 12-row MaxEmergePlus Planter with liquid fertilizer tanks plants no-till corn in soybean residue.

Planting is the most important farm operation in determining how much crop will be available to harvest. In addition to placing the seed in the right environment to germinate, the planter may provide for its feeding and its protection from weeds and insects.

Plants require feeding, just like people, if they are to grow healthy and strong. We need carbohydrates, proteins, and some fats for growth. Plants need nitrogen, phosphate, and potash. Both require adequate water. Nitrogen, in the form of anhydrous ammonia, is often applied by custom applicators prior to planting. Many planters include liquid or dry granular fertilizer attachments to apply starter fertilizer.

Since planters are used for many different crops throughout the United States and parts of Canada, they are offered with a variety of attachments to protect the crop from various pests. Granular **herbicides** (Look in glossary for bold-face words.) to control weeds may be placed in the furrow or in a band as shown right. Spray attachments can also be used to apply liquid herbicides before weeds come up.

A variety of insects attack corn, cotton, and other crops both in the ground and in the standing stalk. Insecticides

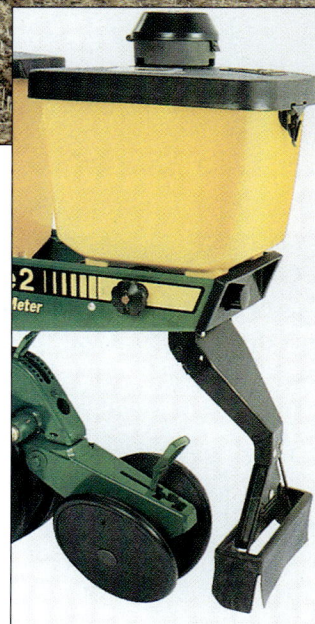

This hopper has chemicals to control weeds in the near side and insects in the far side.

may be applied in the furrow or in a band about 7 inches wide. Insect control chemicals are even more toxic to people than weed control chemicals because insects are more similar to humans. The hopper shown above includes part of the Lid-Fill Closed Handling System developed by John Deere engineers with a chemical company.

The VacuMeter drops seeds accurately to provide uniform spacing in the bottom of the seed furrow.

causes seed to cling to it until the vacuum is cut off and the seeds fall into the seed tube. Individual seeds are deposited in the bottom of the seed furrow with uniform spacing.

Planters offer a variety of row spacings from 15 to 40 inches. The most popular row spacing for corn is 30 inches. If 30,000 seeds are planted per acre, plants will average about 7 inches apart in the row. Corn is the most important crop grown in the United States. The Cornbelt states of Iowa, Illinois, Nebraska, Minnesota, Indiana, and Ohio grow more than two-thirds of the corn in this country. Ontario is the biggest corn producing province in Canada.

Gauge wheels straddling the seed tube are adjusted to provide the desired planting depth to get the seed in contact with moist soil for good germination. Angled closing wheels on either side of the furrow firm moist soil around the seed but leave the soil above the seed loose for easy **emergence** and reduced crusting.

It permits the operator to add insecticides to the hopper without being exposed to the chemicals.

Planters must be versatile machines to plant many different crops such as corn, soybeans, cotton, sorghum, edible beans, peas, peanuts, sugar beets, sunflowers, and vegetables. A chain-drive transmission permits selection of 50 seeding rates by varying the speed the seed meters turn. Seed metering disks are available for each crop. The seed disk, shown above, rotates counter-clockwise through seeds in the planter hopper. Vacuum on the far side of the seed disk

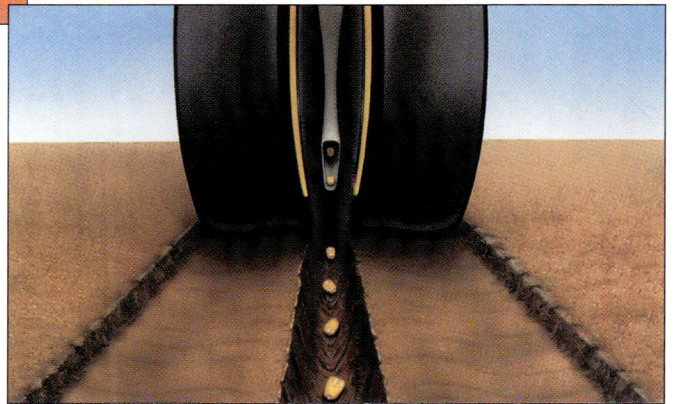

Tru-Vee double-disk openers provide a uniform depth for seed placement.

A single kernel of corn produces an ear with about 600 kernels. If your parents had as many offspring as corn, your sisters and brothers would fill many, many school buses.

What Is Used to Seed Wheat, Oats, and Beans?

A 750 No-Till Drill seeds soybeans in last years corn stubble.

Wheat and oats are much smaller plants than corn, cotton, or sorghum. To get maximum yields, their seeds must be planted closer together in the row and in narrower rows. Grain drills have row spacings of 6 to 10 inches. This distance

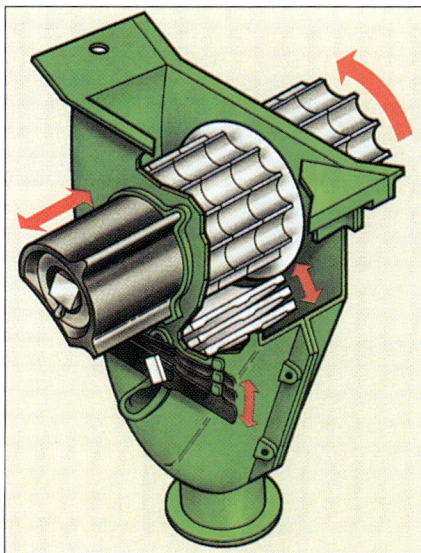

This fluted feedcup uses the latest materials for long life and accurate seeding rates for grain, soybeans, and grass.

is too narrow to cultivate but wheat and oats grow rapidly and normally have limited competition from weeds.

Seeds are metered by fluted feedcups, a basic design that has been used on drills for more than 50 years. The rate of seeding is adjusted by moving the shaft endwise to expose more or less of the rotating flutes. This adjustment controls the volume of seed metered rather than selecting each individual seed like the seed meters do in planters.

Starter fertilizer may be placed in

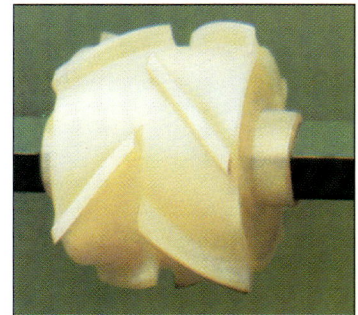

Here is the durable feedroll that pulls fertilizer into the discharge openings.

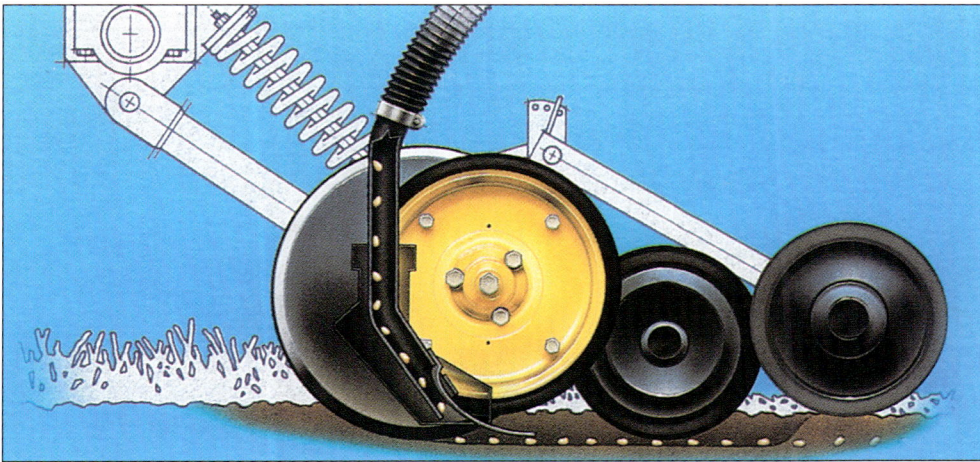

Good seed germination is attained with the 750 No-Till Drill. It uses a down-pressure spring to help the single disk opener slice through residue, a gauge wheel to control furrow depth, a seed tube to place the seed in the bottom of the furrow, a wheel to press the seed into the moist soil, and a closing wheel to cover the seed.

the furrow with the seed or nitrogen fertilizer may be placed at a deeper level between every other row of seeds. Granular fertilizer rates are varied by shifting gears to change the speed of the feedrolls.

Drills were used to seed wheat, oats, barley, rice, and grass even before tractors were popular. In the Midwest, most drills use either single-disk or double-disk openers to open a furrow so seed may be placed at the right depth for germination and growth. In the drier areas of the country, hoe-type openers are popular.

With the development of better weed-control chemicals for no-till farming, it became practical to grow soybeans in rows too narrow to cultivate. John Deere engineers developed the 750 No-Till Drill to seed soybeans and other crops without any prior tillage. Since 1990, this drill has been John Deere's most popular model. Widths of 10, 15, 20, and 30 feet are available to match farm sizes.

Soybeans are planted in the spring, usually after corn planting. Leading soybean growing states include Iowa, Illinois, Minnesota, Indiana, Ohio, and Missouri.

Winter wheat is planted in the fall in Kansas and other states with a similar climate. Spring wheat is planted in the spring in North Dakota

Soybeans are thriving in no-till rye stubble following a spring cutting of hay.

and other northern states. Wheat is one of the most widely grown crops with North Dakota, Kansas, Montana, Washington, Oklahoma, South Dakota, Idaho, Colorado, and Minnesota as major producers. The Canadian provinces of Manitoba, Saskatchewan, and Alberta are also important wheat producers.

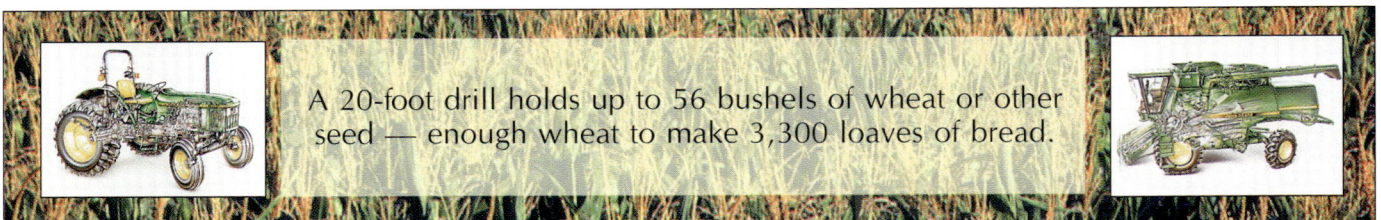

A 20-foot drill holds up to 56 bushels of wheat or other seed — enough wheat to make 3,300 loaves of bread.

What Protects Plants from the Bad Guys?

A Row-Crop Cultivator provides good weed control in 12 rows of corn.

A 400 Rotary Hoe uproots weeds in soybeans.

Weeds and insects are the main enemies of most farm crops. The wheels of a rotary hoe dig out weeds across its entire width, but leave the desired crop undamaged. Corn or soybeans, a few inches high, are not uprooted because they grow from larger seeds that are planted deeper than weed seeds that germinate near the surface. Rotary hoes are operated at about 10 miles per hour for best results. The tines of the rotary hoe spear the ground and, as they come out, explode small clods into the air. These small clods dry out and kill any weeds in them that have germinated.

Rotary hoes are also used some years before the crop has come up. They can break up the crust after a hard rain that might keep some of the desired seedlings from emerging.

Cultivators were, in years past, the main method of weed control for corn, cotton, and soybeans. Today, they remain a good choice to be used in addition to chemical weed control. Cultivators dig deeper than rotary hoes but only between the rows so they do not cut the roots of the crop. They can be adjusted to throw dirt across the row to smother and kill small weeds.

Cultivators can be used for weed control until the crop is more than 2 feet tall before the tractor axles bend the crop over enough to damage it. After that, the leaves of the plant shade the weeds so they don't get the sunlight needed to compete.

A variety of cultivators are used depending on the crop and tillage practices. These include conventional clean tillage, no-till, and variations in between.

Use of chemicals is an essential part of modern farming. However, they need to be applied at the right time, in the right place, and only in the amount needed. This provides the best yields with the minimum environmental problems. In recent years, there has been a shift from farmers doing all their chemical application to more of it being done by custom operators.

Self-propelled sprayers are used for a variety of chemical applications. They can provide

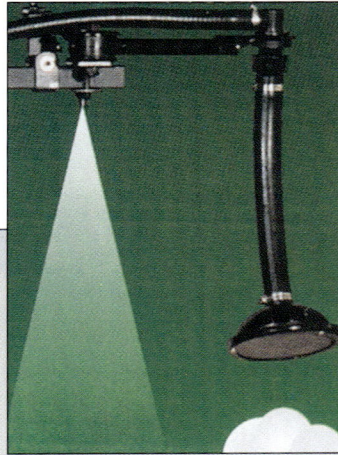

Optional foam markers make a trail of foam dots when rows are not visible to follow. These markers help the operator get full coverage without skips or overlaps.

burn-down to destroy weeds in no-till before planting. Preplant herbicides keep new weeds from germinating. Selective herbicides kill standing weeds and leave the desired crop. Sprayers are also used to apply liquid fertilizers and to apply **defoliants** for cotton. These sprayers have enough clearance under the rear axle to work in tall crops like corn, cotton, cane, and sorghum.

The other major enemy of crops is insects. They may attack plant roots below the ground or the stalk and leaves above the ground. A variety of spray nozzles and locations are available to match the needs for insects or weeds.

Spray attachments are also available on some tillage tools, planters, and cultivators.

A 6500 Hi-Cycle Self-Propelled Sprayer is shown in cotton.

Weed seeds keep lurking in the soil for years, ready to attack. Some ragweeds and pigweeds sprout after 40 years.

Why Is Hay Crushed, Crimped, or Scratched?

A mower/conditioner cuts, conditions, and windrows hay.

Hay is the fourth most important crop grown in the United States. Its area grown and value of production follows corn, soybeans, and wheat. Unlike these other crops, most hay is fed on the farm where it is grown rather than being sold. It is fed to beef cattle, dairy cattle, horses, and sheep.

The big challenge in making good hay is to get it cut at the right maturity, dried enough to store, and baled before it is rained on. In the past, hay was cut by a mower and later windrowed with a rake. Today, most farmers choose to cut, condition, and windrow their hay in a single pass through the field with a mower/conditioner.

Hay type, hay yield, and weather vary widely throughout the country. To handle these different conditions, two methods

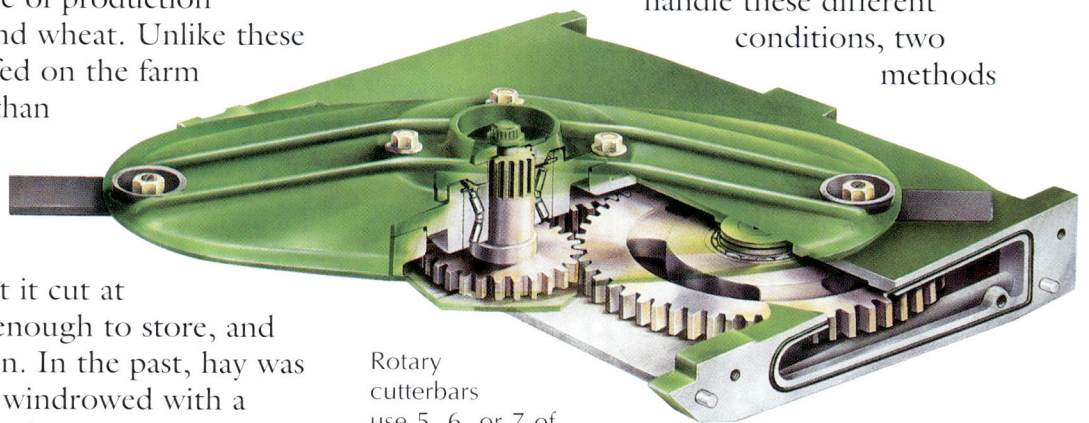

Rotary cutterbars use 5, 6, or 7 of these gearboxes, with flail knives on rotating disks, to cut from 8 feet 2 inches to 11 feet 6 inches wide.

of cutting and two methods of conditioning are provided.

Sicklebars have been used on mowers to cut hay for more than a century and on mower/conditioners for 30 years. They shear the hay like scissors. Many users still prefer this design in alfalfa hay because it takes about half as much power as rotary cutterbars.

Rotary cutterbars have become popular because they can operate at higher travel speeds. They also cut through ant hills, mice nests, and wet dead grass without plugging. The free-swinging knives have a tip speed of 190 miles per hour to cut by impact like lawn mowers. Each disk has a hub that can shear if its rotation is stopped from hitting an obstruction.

Alfalfa has the highest nutritional feed value if it is cut at the first bloom stage and baled at storable moisture with minimum exposure to rain and sun. The time between cutting and baling can be greatly reduced if the stems are opened up to permit rapid drying. Passing alfalfa hay between resilient rolls for conditioning has been used for many years. The rolls are adjusted with enough pressure to crush the stems between them. Stems are also bent or crimped over the corners of the lugs on the rolls. These two actions open up the stem to let it dry down to storable moisture before the leaves become so dry they fall off during baling.

Exposure to the sun is reduced by placing the conditioned alfalfa in windrows, ready to bale. Windrow width can be adjusted for different hay yields and baler requirements.

Urethane conditioning rolls mesh together like gears.

Major alfalfa growing states include California, South Dakota, Wisconsin, Minnesota, Nebraska, Iowa, Idaho, and Michigan.

Feed value of grass-type hay, such as timothy, can also be improved by minimizing the time between cutting and baling. Grass stems dry faster if their waxy surface is removed. V-tine impellers work like fingers to pick up the freshly cut crop from the cutterbar. Wax is removed by the scratching action of the V-tine impellers and by the plants rubbing against the conditioning hood and against each other.

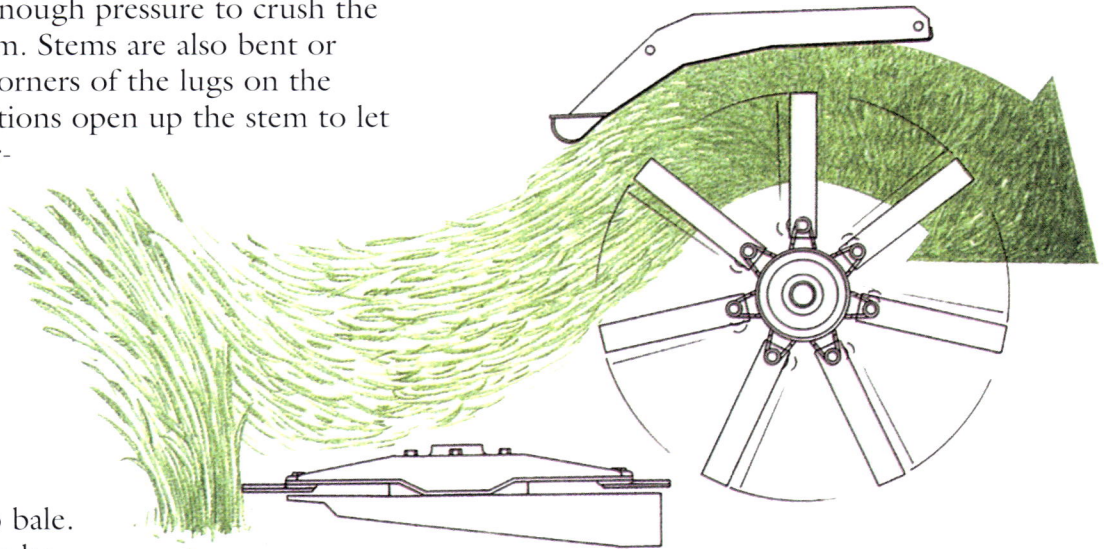

V-tine impellers provide an alternate method of conditioning on rotary cutterbar machines.

About 60 million acres of hay are harvested each year in the United States. If they were all cut by one 920 Rotary MoCo, it would travel enough to go around the world at the equator 800 times cutting alfalfa and 1,200 times cutting other kinds of hay.

Why Are Square Bales Used to Feed Dairy Cattle?

A 338 Square Baler gobbles up a windrow of alfalfa hay.

Dairy cows are probably the hardest working animals on American farms. A typical cow produces about 7 gallons of milk per day. To do this, she must drink about 35 gallons of water. She also eats about 20 pounds of grain and **concentrated feeds** plus about 35 pounds of hay and silage.

Milk production can be increased with proper care and the best feed. Most traditional dairy barns store alfalfa hay in the loft above where the cows are kept, fed, and milked. This keeps the hay out of the rain, snow, and sun. Square bales can be placed in the loft by an elevator but round bales

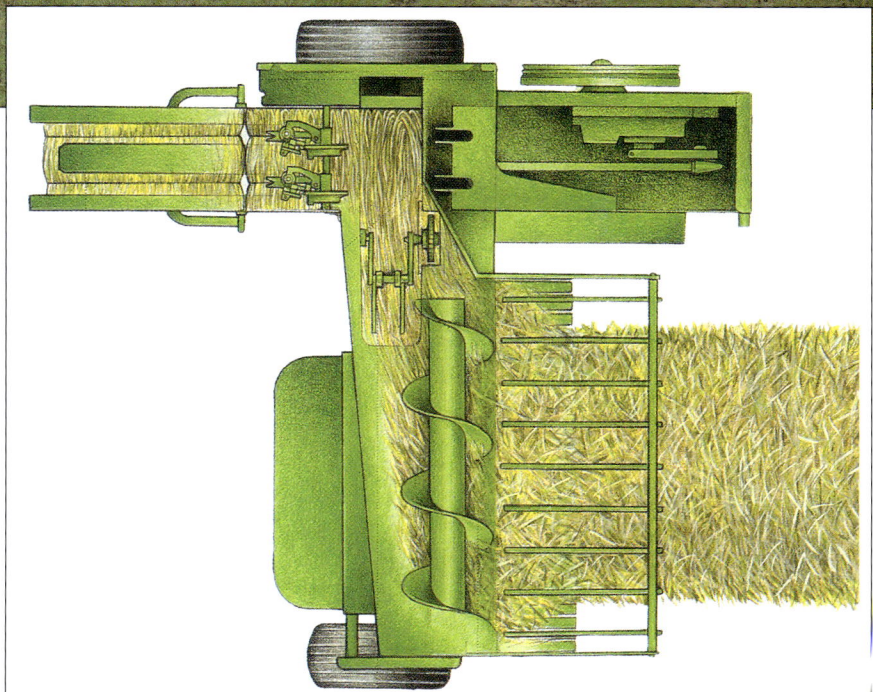

A square baler has many parts that must work together to make good, durable square bales.

are too large and heavy. Square bales are also easier to break apart to feed in mangers.

Some of the top milk producing states are California, Wisconsin, New York, Pennsylvania, Minnesota, Texas, and Michigan.

After the hay in the windrow is dry enough to store, pickup tines lift the hay into the baler gently to save the nutritious alfalfa leaves. An auger continuously feeds the hay across toward the plunger. Twin feeder forks give the hay the final push into the bale chamber. Their action has to be timed so they do not contact the fast moving plunger. The plunger then pushes the hay to the rear of the baler and slices it so the bale can be torn apart easily for feeding. When the new bale is the right length it is tied. Two needles come up from the bottom and the twine is tied by two knotters, all in less than a second. Most bales are 14 inches high, 18 inches wide, and 3 to 4 feet long. They weigh 50 to 80 pounds so you build muscle when you handle them.

After the bale is made, it must be transported to storage. Some bales are simply dropped on the ground to be picked up by some form of bale loader. Many others are pushed up a bale chute extension toward a wagon pulled by the baler. Bales are grabbed off the extension by a person and stacked on the wagon.

The most exciting way to get the bales in the wagon is to throw them in with a 40 Bale Ejector. This unit senses when a bale is ready to throw. Hydraulic cylinders raise the pan under the bale and give it a toss to the rear. The throwing distance can be adjusted by the operator to fill the wagon from the back to the front.

Square bales are stored inside. They are usually placed on an elevator that takes them up into the barn loft. It may be attached to a mow conveyor to distribute them in the loft.

Square balers are also often used to bale the straw left by combines for bedding.

A 40 Bale Ejector throws a bale into a wagon pulled by the baler.

The average person drinks more than 20 bathtubs of milk in his lifetime. How many bathtubs of milk have you finished?

Why Are Round Bales Used to Feed Beef Cattle?

A 566 Round Baler picks up a windrow of alfalfa hay to make a large round bale.

Beef cattle live a life of leisure compared to dairy cows. However, they end up as steaks and hamburgers at a much younger age.

Herds of beef cows and calves are usually larger than dairy herds since they require less labor per animal. There is no need to feed beef cattle individually in mangers. Round bales can be used to feed many beef cattle with minimum handling labor.

Beef cows and calves are grown in every state but do best where there is an abundance of pasture. Some of the top calf producing states are Texas, Missouri, Oklahoma, Nebraska, South Dakota, Montana, Kansas, and Kentucky.

Round bales are used in the winter when pasture is not available.

In contrast, beef steers are fed out in states where corn or sorghum grain is available. Most sorghum is grown in Kansas, Texas, Nebraska, and Missouri. Large feedlots are located in Texas, Kansas, Nebraska, Colorado and Iowa.

Round bales are made something like rolling up a snowball or rerolling paper towels or toilet paper. The tractor and baler straddle a wide windrow of hay. The hay is gently lifted by the pickup tines and fed into the forming chamber. Mating belts roll the bale between them, much like you might roll modeling clay between your two hands.

Making a round bale is shown in three steps.

The small bales weigh about 750 pounds and the larger ones about 2,200 pounds.

Round bales have two distinct advantages. They can be transported, stored, and fed with a tractor doing all the lifting. They can be stored outside because they shed rain. For beef cattle, they are normally stored outside in the field where they were made or brought up to the farmstead.

Dairy farmers that have milking parlors and loose housing also like round bales, but only if they can be stored inside to protect hay quality. Round bales can be transported and stacked three high in a shed with a tractor loader. They are usually fed in a round bale feeder for both beef and dairy cattle. A loader places the bale in the middle of the feeder so cattle can eat on all sides without trampling the hay.

Round balers can also bale straw or corn stalks for bedding.

Hydraulic cylinders control the amount of squeeze the belts put on the hay to make a good, tight bale. When the tractor driver sees the bale is complete he stops the tractor and twine is wrapped around the bale to keep it in a tight roll. This helps the bale shed rain better and makes the bale more durable during handling. After the twine is wrapped, the rear of the baler opens up and the bale rolls out.

Balers are available to make bales from 4 feet wide by 4 feet high to 5 feet wide by 6 feet high.

A round baler discharges a finished bale much like a hen lays an egg.

A 1,000-pound steer gains about 3 pounds per day. If you grew as fast, you would have gone from a 7-pound baby to a 150-pound adult before you were two months old.

Why Is Cattle Feed Cut into Small Bites?

A 3950 **PTO** Forage Harvester chops a heavy crop of corn and blows it into a forage wagon.

Silage is an important part of the ration eaten by dairy cows and beef cattle on feed. Silage goes through a controlled **fermentation** process, much like sauerkraut, between the time it is cut and the time it is fed. This process must be done in the absence of air or it will spoil. Air is excluded by cutting the forage into short pieces at the right moisture content, packing it tight, and minimizing the surfaces exposed to air. Forage harvesters can be adjusted to cut the forage into 1/4- to 1/2-inch lengths.

Most silage is made from alfalfa or corn. Major corn silage producing states include Wisconsin, New York,

John Deere forage harvester cutterheads have 36 or 48 individual knives to cut silage into short pieces.

California, Pennsylvania, Minnesota, Iowa, and Michigan.

Most of the power required by a forage harvester is used by the cutterhead to cut the forage. Sometimes knives on the cutterhead get nicked by rocks or metal. These small individual knives can then be replaced. After much cutting the knives get dull and require more power as they start chewing instead of cutting. The knives can be sharpened by running the cutterhead in reverse and sliding a sharpening stone across their surface. The mating stationary knife is then readjusted closer to the

22

A 6910 Self-Propelled Forage Harvester cuts a windrow of alfalfa and delivers it to a truck.

cutterhead knives as it serves as the other half of the scissors.

Custom operators, beef feedlot operators, and large dairy farmers like self-propelled forage harvesters for their capacity and ability to work in different crops. Two windrow pickups for hay, four row-crop heads for whole-plant corn, and five ear-corn heads are provided to harvest a variety of crops.

Forage harvester headers funnel the crop into the center. As shown in the cutaway, large feedrolls grab the crop from the header and compress it into a mat. This mat is fed to smaller feedrolls that hold it while the mat is cut. A large 48-knife cutterhead then cuts the forage into short lengths. If ear-corn is being harvested, a kernel processor cracks the corn kernels to make them easier for cows to digest. Next, a large blower delivers the crop through the spout to a truck beside, or a wagon behind, the harvester.

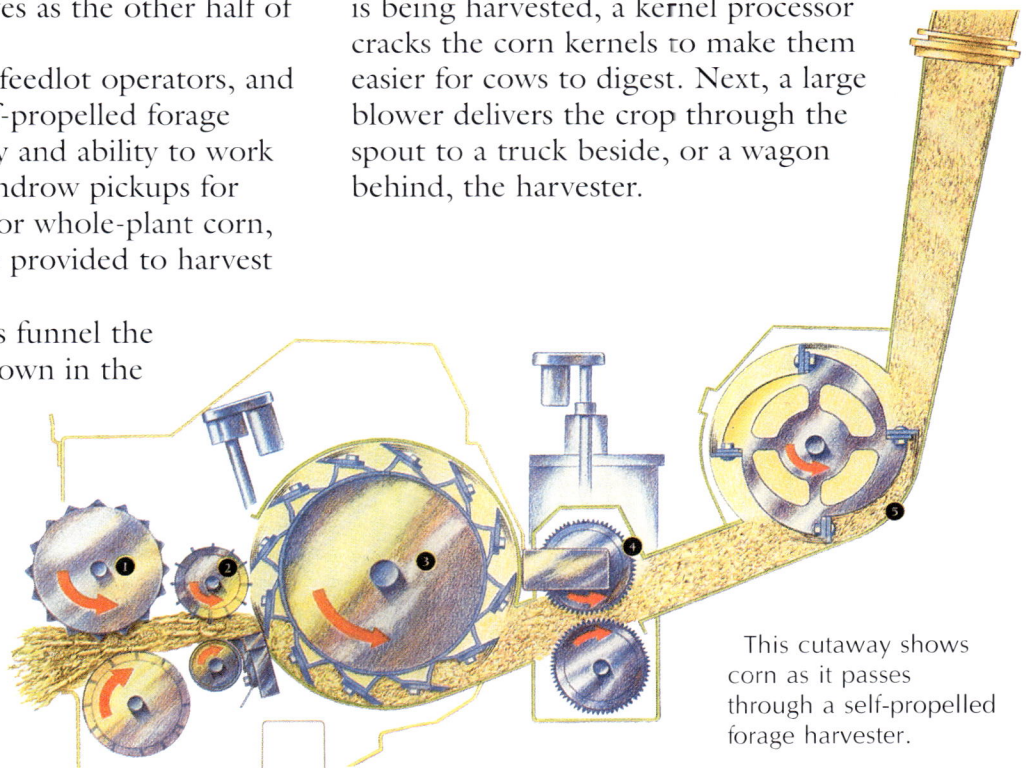

This cutaway shows corn as it passes through a self-propelled forage harvester.

Do your parents ever ask you to eat more slowly? They might be shocked by the 500 bites per second taken by a John Deere forage harvester.

What Do Cattle Eat for Lunch?

A 100 Forage Blower receives forage from a forage wagon and blows it to the top of a vertical silo.

Have you noticed the many tall blue silos or concrete stave silos on farms with cattle? These silos can be as tall as 90 feet. After forage is cut in the field it is brought to a silo by a forage wagon or truck. The forage from the wagon drops onto a rotating feed table that shoots it into the forage blower fan. The fan blows it up the pipe at 90 miles per hour.

Vertical silos were chosen in the past because they kept spoilage to a minimum. The weight of the silage provides its own packing to keep air out.

As both dairy herds and beef feedlots increased in size, horizontal bunker silos with concrete floors and sides became more popular. Horizontal silos provide greater silage capacity at lower cost than vertical silos. Forage wagons can be driven over the silage and unloaded faster. Tractors pack the silage to keep air out. They are covered with sheets of plastic to minimize spoilage of the top surface. Silage can be removed by farm or industrial loaders.

A 135 Total Mixed Ration Machine unloads lunch in a fence-line feed bunk for beef cattle.

Feed for cattle differs considerably from what we eat. Our food pyramid has the most servings of grain in the form of bread, cereal, rice, and pasta. Cattle also eat lots of grain. However, it is usually cracked corn or sorghum rather than processed wheat like we eat. In place of the fruits, vegetables, meat, and milk products we are supposed to eat, they eat much lower cost hay and silage, called roughage. They are able to digest roughage because they have four stomachs while we have only one.

If we had a choice, we might just eat french fries and candy but that is not a healthy diet. Cattle might also "pig out" on grain and not eat their roughage if the feed is not mixed well. Total mixed ration machines are designed to deliver a uniform diet for all the cattle. The farmer determines the best mixture of grain, concentrates, and roughage to use for his dairy cows to produce milk or his feed cattle to produce beef. If baled hay is to be mixed, an optional hay processor is added to the total mixed ration machine.

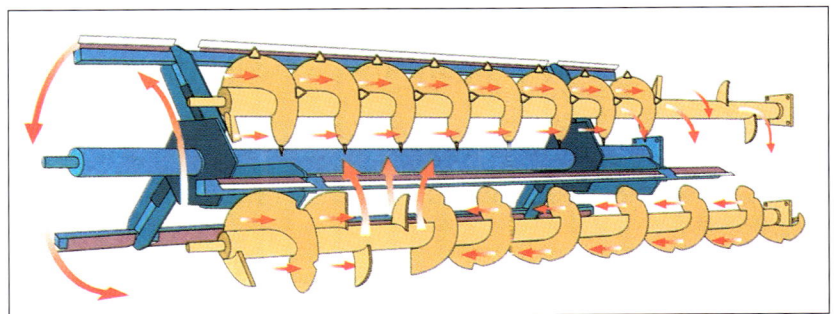

The three-bat rotor and two augers thoroughly mix grain and roughage.

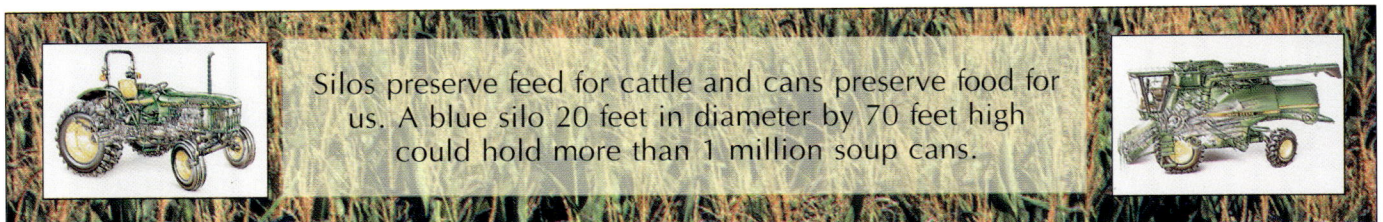

Silos preserve feed for cattle and cans preserve food for us. A blue silo 20 feet in diameter by 70 feet high could hold more than 1 million soup cans.

Why Do Pastures Need a Haircut?

A 709 Rotary Cutter with chain shields cuts and shreds light brush and weeds.

A rotary cutter is an essential tool for many farmers and non-farmers. Pastures provide better grazing for cattle if they are clipped each year to control weeds and brush. Rotary cutters come in widths of 4 to 20 feet. Three duty levels are available, based on the size of brush they can cut.

The 709 Rotary Cutter, shown in set-aside acres, is the heaviest duty cutter John Deere makes. It will cut brush up to 4 inches in diameter or larger than most men's arms. The 709 is also good for cutting stalks of corn, cotton, and sorghum.

Regular-duty rotary cutters are available with one, two, or three rotors. They cut brush up to 2 inches in diameter. They are good for cutting stalks and mowing pastures, orchards, grass waterways, farm yards, and road banks.

Non-farmers find the economy cutters well suited to their many grass mowing needs. Even these cutters handle 1-inch brush without hesitation.

A 2018 Flex-Wing Rotary Cutter cuts corn stalks and distributes the residue evenly with a special attachment.

vertical blade holder shaft.

The splined input shaft to the left of the gearbox has a slip clutch attached to it on many machines to protect the drive line when the cutter hits a stump, large rock, or mound of dirt. Cutters without a slip clutch use a shear bolt for protection. In front of the slip clutch is the shielded, telescoping power shaft with universal joints at each end. This provides the flexibility needed for the cutter to follow the terrain and the tractor to turn corners.

Most hitch-mounted cutters use puncture-proof tires made of recycled car tires. This avoids flat tires from metal or other objects thrown by the blades. Most pull-type cutters use conventional tires for smoother transport.

Rotary cutters cut by impact, the same as home lawn mowers. However, the blades are much heavier. They are attached to the ends of a blade holder and are free to pivot back if they hit a stump or huge rock.

Three choices of blades are provided for different uses. Suction blades pick up grass best from tire tracks. Flat blades work well in cutting brush or stalks. If finer shredding of stalks is desired, a flat blade is used above a suction blade.

The blade holder is attached to the **splined shaft**, shown at the bottom of the gearbox. The bevel gears turn the corner for the power coming from the horizontal PTO shaft to the

Rotary cutters use simple rugged gearboxes. The bevel gears and bearings run in a bath of oil.

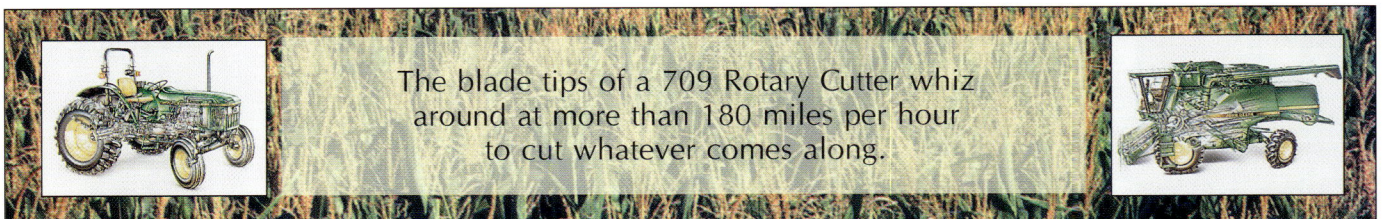

The blade tips of a 709 Rotary Cutter whiz around at more than 180 miles per hour to cut whatever comes along.

What Lifts More than Hercules?

A 640 Loader loads ground feed into a truck. Mechanical self-leveling makes it easy to deliver a full bucket each time.

Farm loaders were originally designed primarily to load cattle manure. Now many farmers have no livestock but still find the loader one of the handiest tools on the farm.

Loaders come with regular-duty buckets that are versatile enough to meet many farm needs. They can be used to scoop up manure, dirt, silage, feed, and snow. Other buckets include economy models with tines for manure, heavy-duty buckets for digging in dirt and gravel, and high-volume buckets for handling silage or snow. Buckets vary from 5 to 8 feet wide, depending on the size of the loader-tractor combination.

Buckets can be exchanged for attachments to handle a variety of jobs. For round bales, there is a choice of a fork with three spears, a bale grapple, or a bale hugger. There is a forklift to move bags of seed or fertilizer on pallets. And a telescoping boom, added to the forklift, provides a high-lift hoist. If the farmer uses more than a single bucket on the loader, he can get a hookup that is fast and requires no tools.

Loaders work better for several chores if the bucket remains level as it is lifted. Some models come with mechanical self-leveling that works both lifting and lowering. Other models offer the option of hydraulic self-leveling during lifting.

A 540 Loader, with mechanical self-leveling, carries a large round bale on a bale fork.

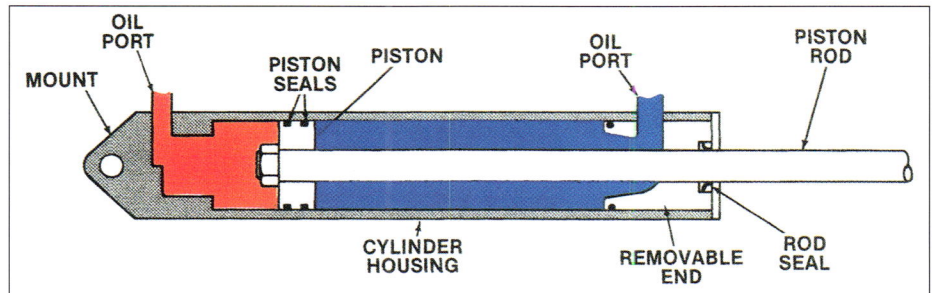

High-pressure oil on the red side of the piston extends the cylinder. High-pressure oil on the blue side retracts the cylinder.

Because loaders are used with many sizes of tractors, 11 models are offered to fit tractors from the small utility models through the large row-crop models. Lift capacity of loaders vary from 1 ton to more than 2 tons. The larger loaders can lift more than the weight of a car or a van. Lift height varies from 9 feet to more than 14 feet.

Since many tractors are used for field work as well as loading, loaders may be mounted and dismounted several times each year. Some loaders can be removed without a wrench. Others require removal of a few bolts from the loader sub-frame, which is left on the tractor.

Four hydraulic cylinders provide the muscle for loader operation. A control valve sends high-pressure oil to the pair of lift cylinders to raise or lower the loader boom. Another valve sends oil to the pair of bucket cylinders to control digging and dumping of the loader bucket.

Loader cylinders can be controlled by use of two selective control valves, present in most tractors. Many farmers choose the optional single-lever joystick control. It operates much like a joystick for computer games.

A 740 Loader can lift 20 muscular men, each weighing 200 pounds, 14 feet in less than 5 seconds.

What Recycles Cattle Manure?

A 740 Loader fills a 785 Hydra-Push Spreader with cattle manure.

Part of the grain and roughage fed to livestock becomes milk or meat but the remainder turns into wastes called manure. Although it is a nuisance where it is deposited, it becomes valuable fertilizer when it is hauled back to fields where crops or pasture are grown. It is recycled as it includes the minerals of nitrogen, phosphate, and potash needed for the growth of plants. It also adds **organic matter** to the soil, which improves its structure and water holding capacity.

Dairy farmers with traditional barns have barn cleaners that load the manure directly into the spreader. This task must be done each day of the year. When it is 20 to 30 degrees below zero, the manure can rapidly freeze to the surfaces of most spreader boxes. But the Hydra-Push Spreader overcomes this problem by having a slick plastic surface on the floor and sides of the spreader.

A Hydra-Push Spreader empties its load much like a garbage truck. The front panel and sliding floor first move to the rear together. Then the floor stops and the panel completes pushing the load into the beater.

The function of all spreaders is to convey a uniform amount of manure to a beater that returns the animal wastes uniformly to the soil to grow future crops. Many farmers let the manure accumulate and haul it in the spring before crops are planted and in the fall after they are harvested. It is also hauled to hay fields and pastures.

Manure must be hauled from beef cattle, dairy cattle, hogs, horses, and sheep. Straw bedding is often used to give the cattle a drier, more comfortable place to lie down. This straw adds to the amount of manure to be hauled. Dirt is also picked up by the loader if the feedlot is not paved. Manure can have lots of liquid in it or it may be dry and hard packed.

To handle these various requirements, different types of spreaders are available. Conventional conveyor chain spreaders work well at temperatures above freezing and cost less then Hydra-Push spreaders. Spreader capacity varies from 177 cubic feet for the smallest conveyor spreader to 341 cubic feet for the largest Hydra-Push Spreader. A hydraulic rear endgate is available on both types of spreaders for handling manure with a reasonable amount of liquid.

V-tank spreaders are able to handle liquid or semi-solid manure. These spreaders seal the liquid in until it is ready to spread. An auger moves the material to a beater that spreads the material sideways over a wide path.

The beaters on a 785 Hydra-Push Spreader chop the manure into small chunks, fling them into the air, and spread them uniformly about 10 feet wide.

Want to be a cowboy? Texas has more than 14,000,000 cattle. Nebraska and South Dakota have more than four times as many cattle as people.

How Are Grains and Seeds Harvested?

A 9500 Combine harvests eight rows of corn with an 893 Corn Head.

An 893 Corn Head finishes a pass through the field.

Combines are the most important tool on the farm after tractors. Farm income depends on harvesting all the crop at the right time before it is damaged by weather or weeds. Combines can be equipped to harvest a variety of crops from large grains like corn to small seeds like alfalfa and bluegrass. The harvest season may start in a dry, dusty wheat field on the 4th of July and end in a muddy corn field at Thanksgiving. Operation in mud can be improved with four-wheel drive. **Flotation** in soft ground can be improved with dual drive wheels.

Farmers work long hours during the harvest season. Some crops are not harvested until the dew evaporates in the morning. Many farmers combine until dark and some work later with lights. Combines are designed to make the farmer's work easier. The cab provides good vision, is heated and air conditioned, and has a good, comfortable seat. Controls and instruments are easy to see and use. Automatic header height is available for soybeans and windrowed wheat. Many farmers unload grain from the combine into trucks or grain carts on-the-go so they don't lose harvest time.

Different types of headers are used with combines, depending on the crop harvested. Headers detach the head of grain from the stalk and feed it into the combine.

For corn, the ears are snapped free of the stalks by snapping rolls. Gathering chains take the ears back to the cross auger, which gathers

A 930 Flexible Platform cuts stalks close to the ground to harvest all the soybeans.

are equipped with pickup reels having nylon fingers that gently lift and move the crop to the cutterbar and then into the cross auger.

The simpler platform, shown above, works well for wheat, sorghum, and barley because they are usually standing. A bat reel feeds the crop evenly into the cutterbar and cross auger.

Other headers are available for special needs. Soybeans shatter easily when dry. A Row-Crop Head shakes the soybeans less so it has even lower losses than flexible platforms. Draper platforms are available for tall rice. Belt pickup platforms are available for windrowed spring wheat grown in the northern states and Canada. Grain platforms vary in width from 12 to 30 feet. Corn heads can harvest 4, 5, 6, 8, or 12 rows.

them in the center and sends them up the feeder house to the threshing cylinder. The points, between the rows on the header, lift cornstalks that have fallen over from the wind or insect damage.

Soybeans present a unique harvesting challenge because the seed pods are located up and down the entire length of the stalk rather than in a head at the top like wheat and sorghum. If the ground is reasonably flat, a rigid platform can cut the plant below the lower pods. However, if the ground is uneven, a cutterbar that flexes across its width can get more of the low growing beans. Automatic header height helps skim the ground to get all the beans.

Selection of platforms and reels is determined by the crops and conditions typically encountered. Soybeans and rice tend to fall down, making it difficult to pick up all the crop. Platforms

Wheat is cut with a 900 Series Rigid Platform.

If all the 9000 Series Combines sold from 1989 through 1997 were lined up, they could cut a swath of wheat more than 100 miles wide.

What Happens Inside Combines?

The many components of a combine work together to thresh the grain, clean it, put all of it in the grain tank, and then deliver it to a truck, grain cart, or wagon.

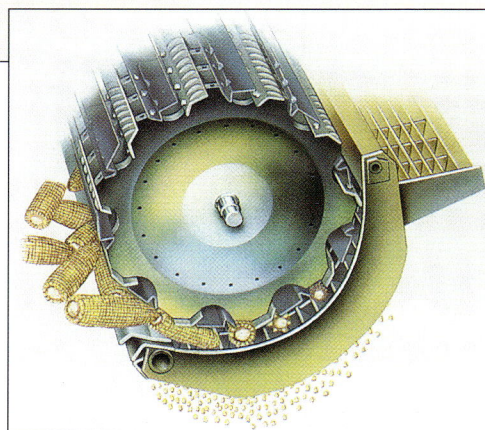

Kernels of corn are removed from their ears as they are rolled between the cylinder and concave.

Grain heads enter the cylinder, shown below the driver's seat, from the header and the feeder house. The cylinder removes the grain from the head by a combination of rubbing and beating as it feeds it over the concave. Most of the grain drops through the concave. But most of the stalks or straw pass to the rear over the shaking straw walkers, shown with saw teeth. Loose grain falls through the straw walkers and joins the grain from the concave. The straw from the walkers can either drop to the ground in a windrow for baling or be chopped and spread in a thin layer.

The grain also has chaff with it. Chaff includes small pieces from the grain head along with broken pieces of leaves and stalks. This mixture drops on the precleaner and moves to the rear over the chaffer. The chaff is blown out the rear and the clean grain drops into an auger where it is elevated to the grain tank. When the grain tank is about full, the long auger at the top of the combine swings out and delivers the grain to a truck or grain cart.

The cylinder, shown above, is rotating counter-clockwise. Its 10 rasp bars contact the ears of corn to shell them as they are fed through. The driver can adjust the severity of threshing action on-the-go from the combine

Grain is cleaned from the chaff by a fan blowing air up through three shaking sieves. The precleaner is in front, followed by the chaffer on top and the sieve on the bottom.

cab. Cylinder speed is adjusted fast enough to get complete threshing and is kept slow enough to avoid cracking the grain. The rasp bars can travel as fast as 75 miles per hour for difficult to thresh crops that do not crack easily. They can be adjusted as slowly at 20 miles per hour for easy to thresh crops that crack easily.

The driver can also adjust concave clearance for severity of the threshing action. Concave clearance is kept close enough to get complete threshing but wide enough to avoid cracking.

Corn is threshed at lower cylinder speeds and with a wide concave clearance so the cobs are not broken. Wheat requires higher cylinder speeds and closer concave clearances than corn. Soybeans are easy to thresh but are also easily cracked when very dry. Cylinder speeds are adjusted much slower and concave clearances somewhat wider than for wheat.

Two principles are used to clean the grain from the chaff. Air from the fan floats and blows out the chaff because it is lighter than the grain. Fan speed can be adjusted from the cab. The second principle used is sizing. The grain is smaller than broken straws or cobs. The lower sieve is set with smaller openings than the upper chaffer as the sieve does the final sizing. Both are adjusted relative to the size of grain being cleaned.

Again, the farmer has to make these adjustments carefully to get clean grain in the grain tank with minimum losses. More air gives cleaner grain but can lose some grain. Closer chaffer and sieve settings give cleaner grain but can lose some grain.

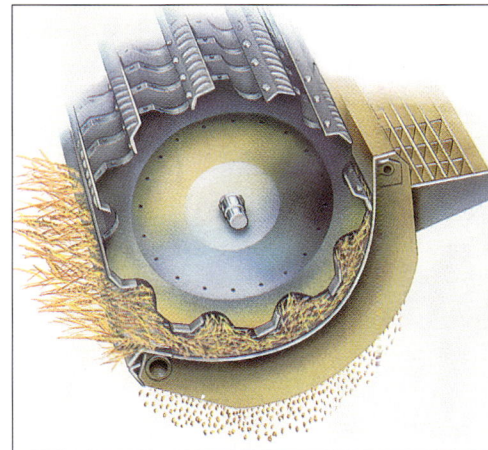

Wheat is removed from its heads as it passes between the rapidly rotating cylinder and the stationery concave.

A 9500 Combine can fill its 204-bushel grain tank with corn in 10 minutes. At that rate, the combine is shelling more than 1,500,000 kernels each minute.

What Crop Is Used to Make Our Clothes?

A 9965 Cotton Picker harvests five rows of cotton.

The most comfortable clothes we wear contain at least 50% cotton. Since cotton absorbs sweat, it feels warmer in the winter and cooler in the summer than synthetic fabrics.

The 9965 Cotton Picker is a very versatile machine with row units that are adjustable for many different row widths. Farmers can easily switch from traditional wide rows to narrower 30-inch rows.

When cotton is about ready to be harvested, a chemical defoliant is sprayed on the plants so the leaves dry and fall off before picking. When the cotton is ripe, the bolls open to expose the fluffy cotton. As the cotton picker row units surround the plant, many finger-like spindles enter it. The rapidly rotating barbed spindles grab the fluffy cotton much like Velcro grabs to fasten clothing. As these spindles revolve around the drum, they contact resilient doffers that remove the cotton from the spindles. Moistener pads apply water to the spindles to keep them from getting gummy. A second drum engages

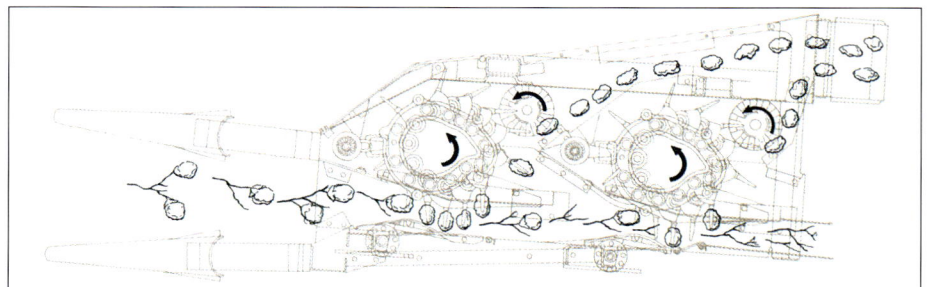

A top view of a row unit shows cotton plants being picked by barbed spindles on front and rear revolving drums.

the plant again and gets the cotton the first drum missed.

From the doffers the cotton is blown up into a huge basket, about the size of a bedroom. When the basket is full, the driver tips it over with hydraulic cylinders to empty it into a **module builder** or into a trailer to be hauled to a **cotton gin**.

Cotton bolls that were not ripe enough to be open during the first picking can be picked later during a second pass through the field.

Some cotton varieties are bred to be harvested in a single pass through the field. Cotton strippers have much simpler row units than cotton pickers because they remove entire bolls rather than just removing the exposed cotton from ripe bolls. A variety of row spacings are easily obtained on cotton strippers with a maximum of six rows in narrow-row cotton.

The row units have some similarity to corn head row units. However, the stripper row units use inclined brush rolls that sweep up to remove the cotton bolls rather than snapping rolls that pull corn stalks down. The harvested bolls are augered to the rear where they drop into the cross auger that brings them to the middle. The open bolls are blown up the air duct to the basket while the closed green bolls drop back on the ground. When the basket is full, it is dumped sideways into a module builder or trailer.

A 7455 Cotton Stripper harvests six rows, stripping them clean of cotton bolls in a single pass.

If the farmer wants cleaner cotton, he can divert the cotton from the row units through an optional saw-type cleaner to remove most of the trash from the cotton. The cleaned cotton is then blown into the large basket which holds enough cotton to make three bales.

Major cotton producing states include Texas, California, Mississippi, Georgia, Arkansas, and Louisiana.

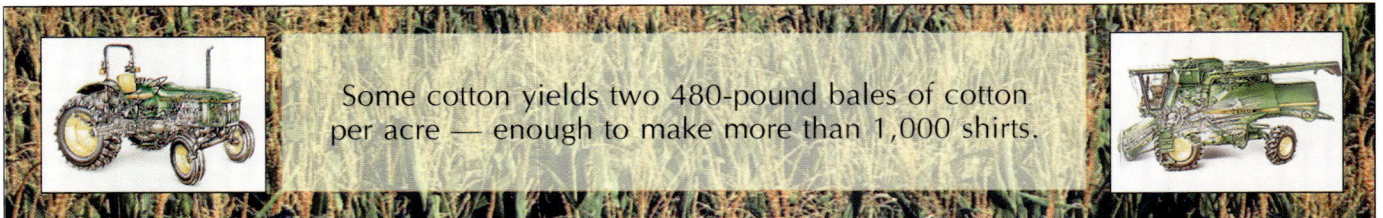

Some cotton yields two 480-pound bales of cotton per acre — enough to make more than 1,000 shirts.

What Has More Power than a Hot Sports Car?

This cutaway shows the parts inside the 4-cylinder **turbocharged** engine used in 6000 Series Tractors.

▽ This cutaway shows the valves that control the breathing of this series engine.

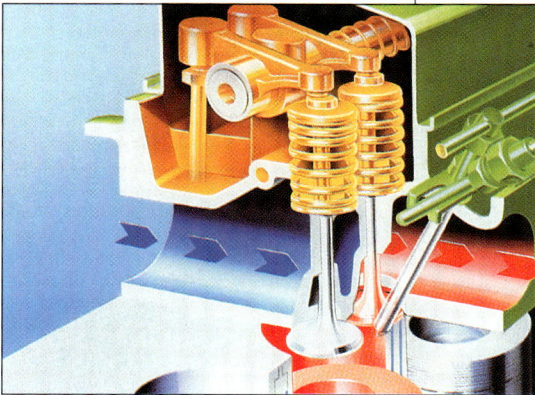

John Deere tractors and self-propelled equipment are powered by three different series of diesel engines. The most widely used engine is the smallest of the three. It is made in Dubuque, Iowa, and in three other countries.

The engine covers a wide range of powers as it is offered with 3, 4, or 6 cylinders. A

3-cylinder, 2.9-liter, **naturally aspirated** engine provides 40 PTO horsepower for the 5200 Tractor. A 6-cylinder, 6.8-liter, turbocharged engine provides 115 PTO horsepower for the 7610 Tractor.

This engine is also widely used in self-propelled equipment. The 4-cylinder version powers the 6000 Series Sprayers and the 4890 Windrower. The 6-cylinder version powers the 4700 Sprayer, the 9400 Combine, and the 7455 Cotton Stripper.

Intake air, shown to the left in blue, enters the engine from the left and enters the cylinder through the left valve when the piston goes down. When the piston returns to the top, the pencil-shaped fuel injector nozzle sprays fuel into

This 6-cylinder turbocharged diesel engine is from the series found in most John Deere row-crop tractors in the United States.

provides a 50% **torque** rise. This rise in torque, as the engine receives a heavy load, permits it to keep the unit moving without shifting gears.

All engines in this series shown below are turbocharged and air-to-air after-cooled. The 10.5-liter size provides 310 gross engine horsepower in the 9200 Tractor. A muscular 12.5-liter version provides 425 gross engine horsepower in the 9400 Tractor. This is more power than hot sports cars.

This modern engine features a cam in the head that operates the four valves per cylinder and the unit fuel injectors. The injectors are electronically controlled for fuel efficiency, clean burning, and good torque rise.

the compressed, heated air. The fuel-air mixture explodes, driving the piston down to provide power. As the piston comes up again, the exhaust gases shown in red leave through the exhaust valve and out the right side of the engine.

All engines in this series shown above are made at Waterloo, Iowa, as turbocharged 6-cylinder units. All current tractors using this engine have 8.1-liter **displacement**. However, a range of powers is offered by varying fuel and air delivery. The engine provides 130 PTO horsepower in the 7710 Tractor. **Air-to-air aftercooling** is used in the 8000 Series Tractors. The 8400 Tractor features 225 PTO horsepower. A similar engine is used in the four-wheel-drive 9100 Tractor.

In self-propelled units, turbocharged 6-cylinder versions are used in the 6710 Forage Harvester, 9500 and 9600 Combines, and the 9965 Cotton Picker.

These engines feature an in-line fuel injection pump with electronic engine control that

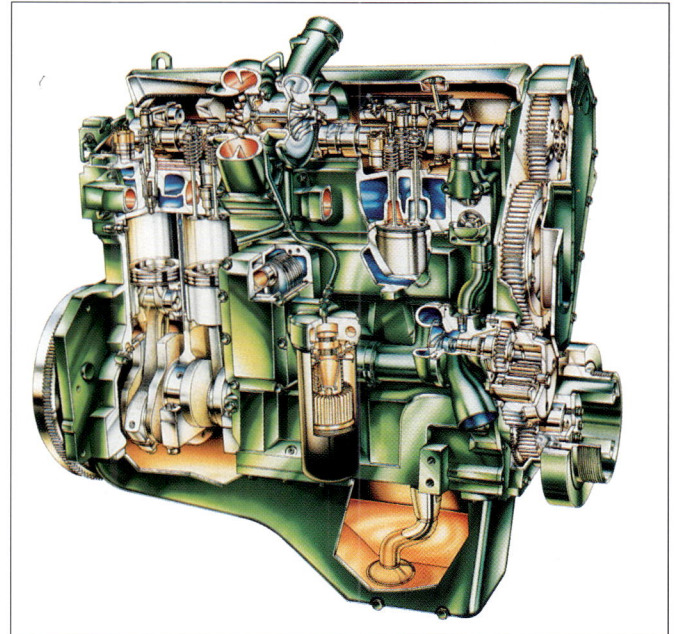

This is the latest and largest 6-cylinder engine made at Waterloo for four-wheel-drive tractors.

A 7810 Tractor engine breathes as much air in a minute as an adult does in a day.

Where Can We Get More Muscles and Brains?

All controls on 8000 Series Tractors are within easy reach. The most used ones are on the right armrest of the seat.

Engineers continually improve tractors and self-propelled equipment to make the driver's work easier and more productive. More than 60 years ago, a hydraulic lift was added to lift equipment mounted on tractors. A decade later, hydraulic cylinders were used to lift pull-type implements.

While these took the muscle work out of implement lifting, tractors were getting bigger and harder to steer. In 1954, John Deere offered power steering on tractors. This advancement was soon followed by power brakes and a power differential lock.

Today's wide, fast, and productive implements would not be practical if hydraulics were not used to control both the tractor and the implements.

Hydraulic systems include a pump to provide oil under pressure, control valves to determine where the oil goes, connecting lines, and hydraulic cylinders to do the work. A cutaway hydraulic cylinder is shown on page 29. Cylinders to control the tractor, such as power steering, are relatively small. Loader cylinders, 3-point hitch cylinders,

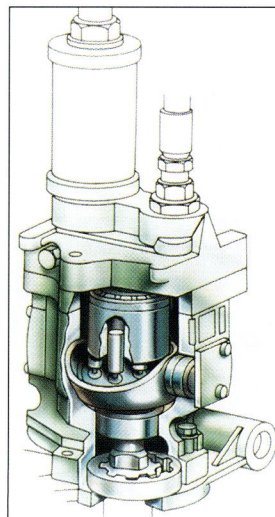

The main hydraulic pump on 7000 Series Tractors uses multiple pistons to power most hydraulic controls.

Hydraulic circuits are concentrated at the rear of the 7000 Series Tractors, with the main pump located below the seat.

and those used to fold wide implements are larger diameter and longer because more work must be done.

In addition to the previously mentioned uses for hydraulics, they are used to operate clutches for the main transmission, the PTO, and the mechanical-front-wheel drive. Hydraulics are also used to control clutches and brakes in optional power-shift transmissions. Hydraulics are further used to power hydraulic motors on planters and other implements.

While hydraulics have reduced the muscle work of operating equipment, it is electronics that have multiplied the driver's brain power. Electronic fuel delivery on engines make them easier to start, burn cleaner, use less fuel, and provide added power under heavy loads. Electronic control of power-shift transmissions provides rapid and smooth shifting. On the 8000 and 9000 Series Tractors, a small lever on the armrest is used to shift gears.

Electronics have teamed up with hydraulics to provide simpler, more precise, fingertip control of implements.

Electronics monitor more tractor functions than before. Optional radar lets the driver know if his wheels are slipping too much and wasting fuel. Planter monitors tell the driver if all rows are being planted correctly. Combine monitors tell the driver how he has adjusted the combine and how it is functioning.

Last, but not least, electronics provide the driver with a radio or cassette tape to listen to while he works in a quiet, comfortable, temperature-controlled cab.

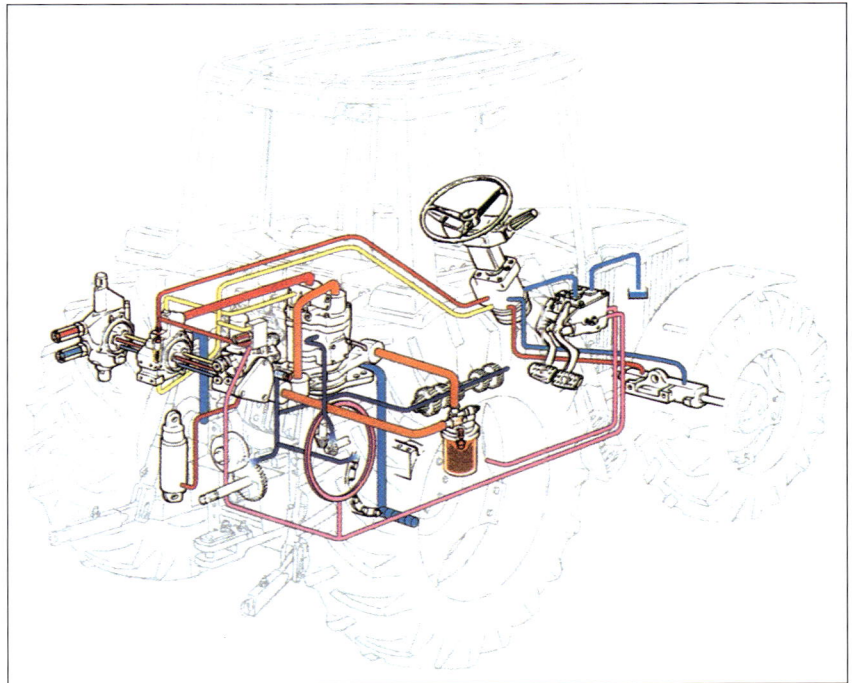

The 3-point hitch on 7000 Series Tractors uses electronics to control the hydraulic lift.

Most new John Deere tractors can provide 2,900-pounds-per-square-inch hydraulic pressure. That is more than 30 times the pressure of water at the faucet or hose at most homes.

How Does Power Get from the Engine to the Ground?

This 7800 Tractor is shown with **MFWD** and 16-speed PowrQuad transmission.

Farmers like their implements to cover as much ground as their tractors will handle. This requires a good selection of speeds that are easy to change on-the-go to match field conditions. Properly loaded engines are more fuel efficient.

A cutaway of a complete 5300 Tractor is shown on the back cover. A clutch on the flywheel at the rear of the engine transmits power from the engine to the transmission, shown above. Travel speeds are determined by the combination of gears engaged in the transmission by the gearshift lever. At the rear, bevel gears turn the corner to drive the differential. The differential gives equal power to both rear axles but permits one wheel to travel faster than the other during turns. A planetary final drive reduces the speed and increases the torque to the rear drive wheels.

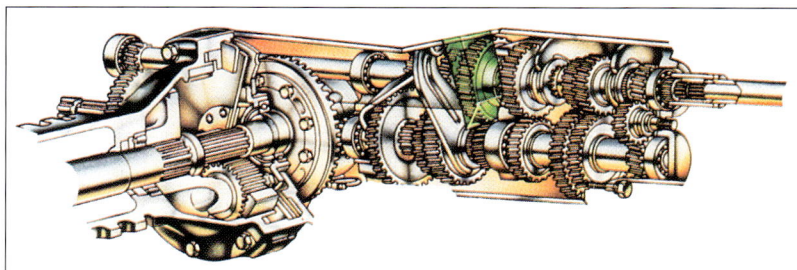

The 5000 Series Tractors have a simple transmission with 9 forward and 3 reverse speeds.

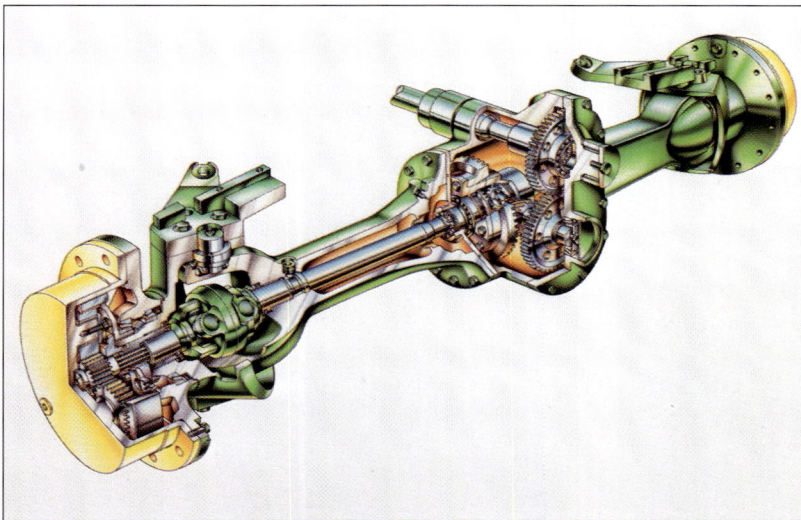

The mechanical front-wheel drive shown is a desirable option on 8000 Series Tractors to fully utilize their high horsepower.

John Deere engineers overcame one of the biggest challenges in MFWD design on both the 7000 and 8000 Series Tractors. These high-horsepower tractors can make tight turns at the end of the field even with the wheel tread set at 60 inches for narrow-row corn cultivation.

Tires are the final connection between the engine and the ground. They deliver the tractor's power for tillage, planting, loading, and several other operations. Some of the tractor's power is used by the PTO when making hay or forage.

Tires with a large footprint (area in contact with the ground) save fuel, get more work done, and reduce soil compaction. Footprint size can be increased by using larger diameter tires, wider tires, and lower inflation pressures.

The standard transmission on 6000 Series Tractors has 12 forward speeds and 4 reverse speeds. Gears may be shifted on-the-go with clutching.

The 16-speed PowrQuad transmission is also available on all 6000 and 7000 Series Tractors. It provides four power shifts, without clutching, in each of four ranges. It is especially good for loader operation because the direction of the tractor can be changed smoothly without clutching.

A 19-speed power shift transmission is an option on the three largest 7000 Series Tractors. A 16-speed power shift transmission is used on all 8000 Series Tractors.

With mechanical front-wheel drive (MFWD), some of the tractor power goes through bevel gears to drive the front axle differential. Universal joints in the axles permit the front wheels to turn corners. The final planetary gear reduction is in the wheel hub.

MFWD is a popular option on all tractors for field work. Its larger tires also provide better flotation in soft ground. MFWD is especially useful for loader operation in slippery conditions. The added weight the loader puts on the front axle can be used for better traction.

Properly selected and inflated radial tractor tires will have bulges or cheeks as shown above.

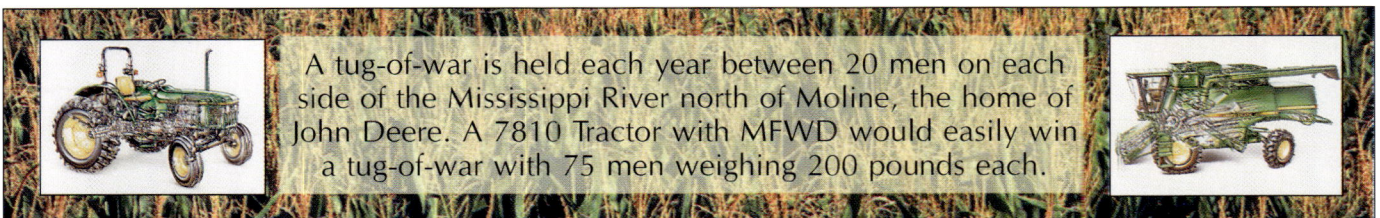

A tug-of-war is held each year between 20 men on each side of the Mississippi River north of Moline, the home of John Deere. A 7810 Tractor with MFWD would easily win a tug-of-war with 75 men weighing 200 pounds each.

What Is a Yield Map?

A yield map shows interesting variations in a soybean field with an average yield of 34 bushels per acre.

**John Deere West Farms
Valley East Field**

Field Size: 60.00 Acres

Harvested Acres: 59.18
Yield: 34.25 BPA
Moisture: 13.5 %
Processing Year: 1996

GREENSTAR

Soybean Yields

45 and greater	30 to 34
40 to 44	25 to 29
35 to 39	less than 25

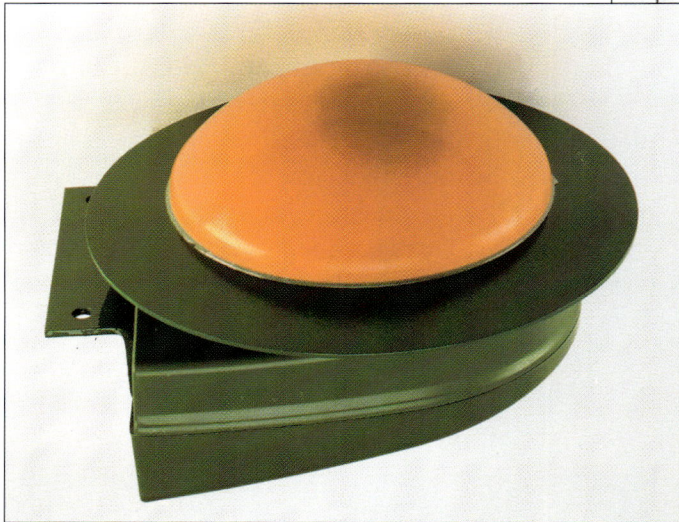

A Position Receiver is located at the top of the combine to get signals from Global Positioning Satellites.

Many new John Deere combines are equipped with the optional GreenStar precision farming system. The purpose of the system is to increase farm income through increasing yields and reducing costs of producing the crop. The first step is to break large fields down into much smaller areas showing their crop yields. The second, and more difficult step, is to analyze why some parts of the field have lower yields and what can be done about it. Is it because the soil type is different or the ground too wet, too dry, too steep, or too compacted? Are yields reduced by competition from weeds?

Can input costs be reduced by varying the amount and type of fertilizer applied, changing seeding rates, or targeting herbicide application to just the problem areas?

The position receiver locates the combine or other equipment in the field, accurately within 7 feet. This information provides the location for producing the original yield map. Soil testing is done in many locations within the field to determine how lime and fertilizer rates need to be varied. A position receiver is also used later to vary application rates of seed, fertilizer, or herbicide on-the-go within the field.

Grain, leaving the elevator, strikes a movable plate that varies its signal according to how much grain is flowing. This provides information required to draw the yield map.

Moisture content of the grain must also be determined to convert the yield to dry bushels

MASS FLOW SENSOR

CLEAN GRAIN ELEVATOR

LOADING AUGER

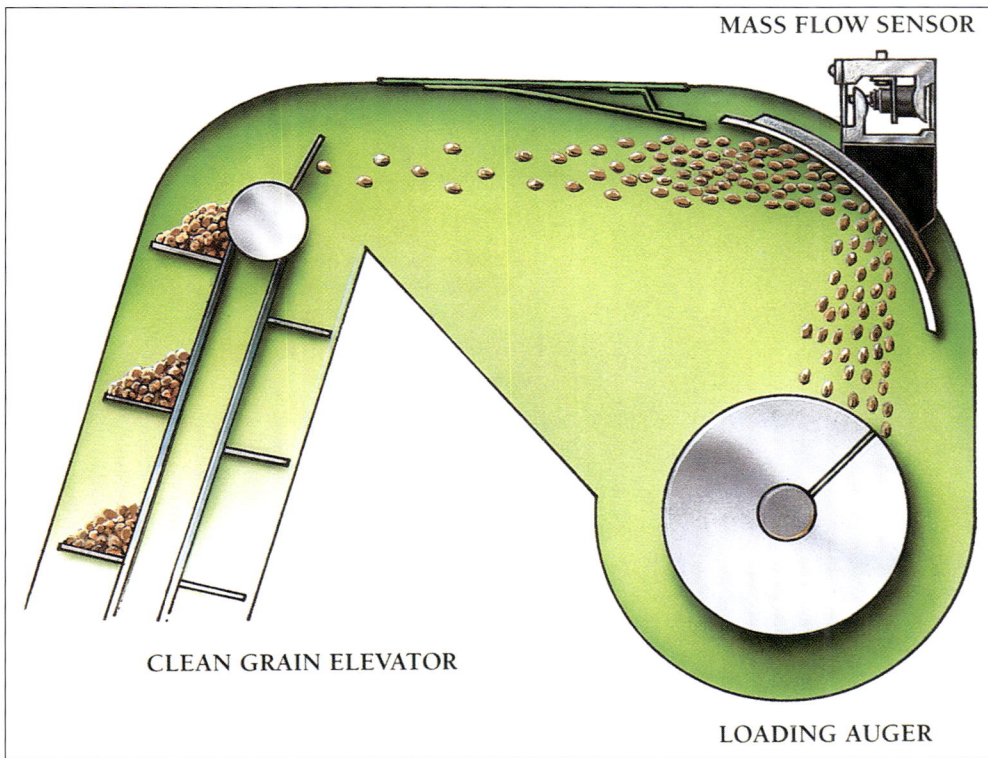

The yield sensor is located near the discharge of the clean grain elevator.

per acre. The farmer also wants to know moisture content to determine if the grain must be dried for safe storage. A moisture sensor is attached to the side of the clean grain elevator. Some of the grain from the elevator is sent through the sensor where its moisture is detected between two metal plates.

When combining, the GreenStar display provides on-the go information on

The GreenStar Display is used on the combine during harvest and other equipment during planting.

wet and dry yield and grain moisture. Keys on the display permit the farmer to select other menus to receive the harvest information he wants. The operator can also flag areas to mark weed patches or wet spots to be printed on yield maps.

At the end of the day, the farmer removes a PC Card from the combine processor and transfers the information to his home computer. The computer can then be used to print out the colored-coded yield maps.

The same GreenStar display is moved to other equipment and used with other menus to control crop input rates. These include the application of fertilizer and herbicides as well as planting rates. When planting, one menu uses the display to monitor planter functions while another menu permits the driver to select up to six different planting rates on-the-go. In the future, variations in application rates can be map based with the use of the position receiver.

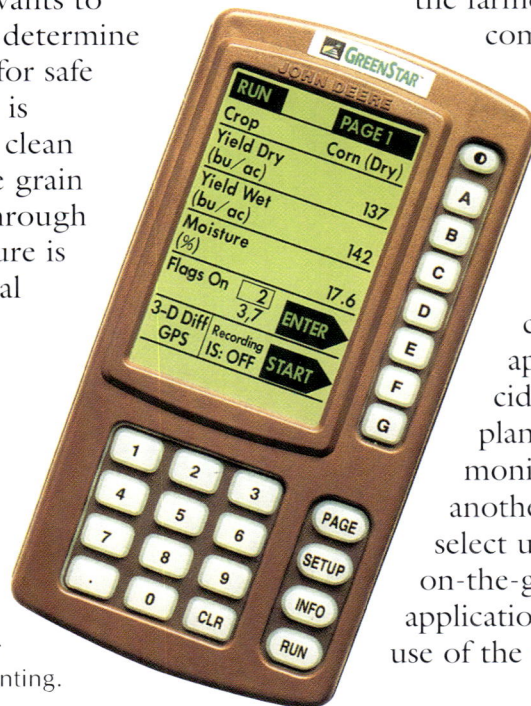

Do you want to become an agricultural engineer or an agronomist to help farmers increase their income from yield maps and precision farming?

45

Glossary

Air-to-air aftercooling — A system used to cool the air from the turbocharger to get more air into the engine for increased power.

Concentrated feeds — High-value livestock feeds, purchased by the farmer, that include protein, minerals, and sometimes medicines.

Cotton gin — A processing plant that separates the cotton fibers from the seed and trash before baling the cotton.

Defoliants — Chemical sprays that cause plant leaves to dry and fall off.

Displacement — A measure of engine size determined by cylinder area, stroke length, and number of cylinders. John Deere makes engines from 2.9 to 12.5 liters displacement. Most car engines are 1.6 to 3.8 liters.

Emergence — The act of a young plant coming through the soil surface.

Fermentation — A process for saving the high nutritional value of a crop for feeding at a later date. Alfalfa and corn have a very high moisture content when they have their highest feed value. If simply piled, they would mold and lose their feed value. Properly stored in a silo in the absence of air, favorable bacteria causes the silage to ferment and keep its feed value.

Flotation — The ability to stay on the surface of the ground rather than cutting ruts.

Herbicides — Chemicals used to kill weeds.

MFWD — Mechanical front-wheel drive for tractors, similar to four-wheel drive on pickups.

Module builder — A large machine, with four walls, to receive and pack cotton for storage as a stack in the field before being sent to the cotton gin.

Naturally aspirated — The basic system for an engine to draw its air through an air cleaner and the intake manifold.

Organic matter — Material resulting from the decaying of plants or animals.

PTO — A power take-off is a rotating shaft on the tractor used to power implements.

Splined shaft — A shaft that has teeth on it, like a gear, used to transmit power in tractors and implements.

Torque — The rotational force of a shaft.

Turbocharged — An engine that has more air forced into the cylinders so they can efficiently burn more fuel to produce more power. A turbocharger gets its energy from the exhaust gases to drive the turbine that compresses the air entering the engine.

Index

Other Information on John Deere

http://www.deere.com — Visit John Deere's home page to learn more about equipment and the company that makes it. You can also learn about Career Partnership, a joint program between selected colleges and John Deere.

1-800-522-7448 — You can call this number or write Cindy Calloway, John Deere Publishing, Deere & Company, John Deere Road, Moline, IL 61265 to order publications. High schools and colleges use educational materials from Fundamentals of Service, Fundamentals of Machine Operation, Farm Business Management, and the new Agricultural Primer Series. This is also the source for Operator's Manuals, Parts Catalogs, and Technical Manuals. Any of the above can also be ordered through your local John Deere dealer.

John Deere Pavilion — The Pavilion is the place to see new and old John Deere farm equipment. It also has interactive displays on farming. It is located at 1400 River Drive, Moline, Illinois, near the site of John Deere's first plow factory in Moline.

Factory Tours — There are several locations where you can see John Deere equipment made, with your parents, if you are 12 years or older. Look at the World Wide Web for factory locations and phone numbers.

About the Author

Roy Harrington was always fascinated by how things worked as he grew up on a livestock and grain farm. He worked more than 30 years as an engineer planning and developing John Deere farm equipment. Roy is co-author of *John Deere Tractors and Equipment 1960-1990*, a best seller farm equipment history book. He is also author of two children's books, *A Tractor Goes Farming* and *Grandpa's John Deere Tractors*. His five grandchildren are fascinated by the three tractors he drives.

About ASAE — The American Society of Agricultural Engineers

ASAE is a technical and professional organization committed to improving agriculture through engineering. Many of our 8,000 members in the United States, Canada, and more than 100 other countries are engineering professionals actively involved in designing the farm equipment that continues to help the world's farmers feed the growing population. We're proud of the triumphs of the agricultural and equipment industry.

High Calorie
SHAKES

Recipes for older
adults experiencing
unintended weight loss.

The
Geriatric
Dietitian

High Calorie
RECIPES

Katie M. Dodd, MS, RDN, CSG, LD, FAND
Board Certified Specialist in Geriatric Nutrition

TABLE OF CONTENTS

NOTE: This book was originally written for caregivers. They are often the ones preparing the foods. However, if you are making these shakes for yourself, I'm so glad you are here. These recipes are intended to help you stop weight loss, gain weight, and protect your muscle! I hope you enjoy these recipes!

AUTHOR'S MESSAGE

Hi! My name is Katie Dodd. I am a geriatric dietitian. I am here to help you provide delicious high calorie shakes to help stop unintended weight loss in older adults!

Why is unintended weight loss such a big deal in older adults? Did you know that unintentionally losing just 5% of body weight in just 30 days increases the risk of death 10-fold??

When we lose weight unintentionally, we lose muscle. And it is my goal that we protect muscle in older adults to promote independence and quality of life! We shouldn't wait until someone becomes malnourished. When weight starts dropping, we should intervene ASAP!

This cookbook is targeted to older adults. However, anyone in need of a high calorie diet can benefit from these recipes. As an added bonus, I am including my Weight Gain SECRETS at the end of this book.

I hope these shakes are enjoyed and thank you for taking care of the older adult in your life.

xoxo,
Katie Dodd

Here is my disclaimer: I am a dietitian, but I am not your dietitian. This book includes information for the general public and should not be considered medical advice. Please work with your healthcare team to make sure you are getting the individualized care you need! Also note calorie counts are estimates and may vary by individual products and brands used.

WHY HIGH CALORIE?

Hold up. High calorie? Aren't a ton of calories and lots of fat bad for you??

For some people, using high calorie recipes can be a huge shift. We are so used to focusing on "healthy foods" high in vitamins and minerals. As a dietitian I can attest to the trans-formative power of food.

However, when an older adult is unintentionally losing weight, they are losing muscle (read= losing independence). They are becoming malnourished. This increases their risk of hospitalization, falls and even early death. This is a big deal.

Providing high calorie foods can stop weight loss.

Foods high in fat have the most calories per volume than any other nutrient. When unintended weight loss occurs, stopping it becomes the priority.

Think about it, we eat healthy foods to prevent the long-term complications of disease and illness... but if an older adult loses weight and muscle, if they fall, are hospitalized, or if they die as a result... what was the point?

When facing malnutrition and the risk of losing muscle and even death, calories and stopping weight loss takes priority over eating "nutritious foods".

Now keep in mind, the goal isn't to stay on a high calorie diet forever. The goal is just to stop the weight loss. And to promote weight gain as needed. But there will be cases where people are on a high calorie diet for an extended period of time.

When it comes to providing a high calorie diet, we also want to provide delicious, tasty foods people will eat. I've found even when an older adult can't take another bite of food, they can still sip on a high calorie beverage.

High calorie shakes can help stop weight loss! These high calorie shake recipes can help stop unintended weight loss in older adults. It makes it easier to promote independence and quality of life. And who doesn't love a delicious, tasty shake? Enjoy!

4

HIGH CALORIE FOODS

Here are some foods that are typically high in calories. Be sure to read the nutrition facts label for calorie content as different products/ brands may vary.

DAIRY

FATS

SWEETS

OTHERS

DAIRY

Whole milk
Half and half
Buttermilk
Chocolate milk
Powdered milk
Greek yogurt
Cottage cheese
Cream cheese
Ice-cream

FATS

Avocado
Olive oil
Butter
Nuts
Nut butters
Soft spread margarine
Egg yolks

SWEETS

Syrup
Honey
Sugar
Agave
Jam
Hazelnut spread
Flavored syrups

OTHERS

Dried fruit
Coconut
Pre-made nutrition shakes
Protein Powder
Flax seed

SHAKE BREAKDOWN

Most shakes come from a simple recipe:

- a liquid base
- a tasty flavor
- some high calorie add-ins

#1 Start with a liquid base.

- Whole milk
- Half- and- half
- Buttermilk
- Heavy whipping cream
- Pre-made nutrition drinks
- High calorie milk substitute

#2 Add a tasty flavor!

- Chocolate syrup
- Caramel syrup
- Strawberry syrup
- Fruit jam or jelly
- Butterscotch syrup
- Hot fudge sauce
- Hazelnut spread
- Honey

#3 Top-off w/ high calorie add-ins.

- Ice-cream
- Peanut butter
- Coconut cream
- Powdered milk
- Nut butters
- Greek yogurt
- Banana
- Protein powder

#4 Blend until smooth!

Experiment with the amount of each ingredient, add more liquid or solids as needed.

TOOLS OF THE TRADE

Measuring Cups

Use this for liquids

Use these for solid ingredients

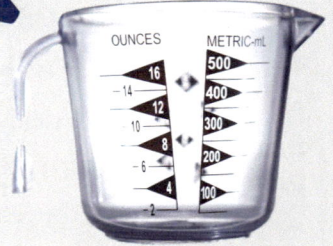

Good Blender

The secret to a good shake is a powerful blender! You can use:

- Full-sized blender
- Single-serving blender
- Commercial milkshake blender

Cups & Straws

Milk shakes are best enjoyed with a straw, though it's not required. You may opt for a cup with lid to prevent spills. Others prefer a shake in a tall, cold glass.

Storage for Extras

Most of these recipes will fit in a **20 oz. cup**. If you make extra or this serving size is too much for one sitting, be sure to store the leftovers in a sealed container in the fridge to prevent food borne illness.

TOP SECRET TIPS

Temperature

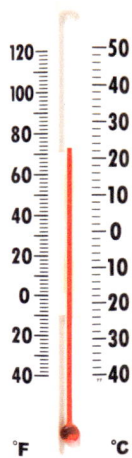

Most people expect and prefer a cold milkshake. But keep in mind not everyone does. Some people are sensitive to cold drinks. They may prefer a shake prepared at room temperature (or what has melted in the fridge). Others may want their shakes extra cold. Be sure to ask about individual preferences.

High Calorie "Ice"

Some people add ice to shakes to make them extra cold. BUT, ice is made of water. Water has zero calories. We want to maximize every ounce of shake with calories. Freeze half and half or whipping cream in ice cube trays. Then use this high calorie "ice" in your shakes!

"Fortified" Milk

"Fortified" milk is a trick of the trade for maximizing calories in milk. You do this by adding powdered milk to liquid milk. Ex. 1 cup whole milk with 1/4 cup of dry milk powder. You can serve a glass of milk this way for an extra boost. Or you can use "fortified" milk in any of these milkshake recipes!

Recipes in Bulk

As a caregiver you are busy. You may want to make recipes in bulk. Just decide how many servings you want to make and multiply the ingredients by this number to make a recipe in bulk. Store the leftovers in the fridge. If a cold shake is preferred, store half in the fridge and half in the freezer. Then blend together when ready to serve!

Food Safety

Older adults often have compromised immune systems. Proper food handling is imperative to prevent food borne illness. Always be sure to use ingredients stored at proper temperatures. And remember that rinsing a blender with water only is not enough. We cannot "see" germs with our eyes. Clean the blender with soap and water in between each use.

TOP SECRET TIPS

Timing

It is best to offer high calorie shakes **in-between** meals. This way they are getting good nutrition from a meal and the shake isn't just a "meal replacement". Don't provide shakes too close to a meals or they may be too full to eat. Everyone is different, so experiment with timing!

Display

We eat with our eyes! The way something looks has an impact on our desire to eat it. Pay attention to the details. When appetite is poor, this makes a big difference. Try whipped cream or garnish to improve the display of your shakes.

Thickness Preference

Most of the shake recipes in this book are made **thinner** for older adults. This is to reduce the amount of effort to suck the shake up a straw. Thin shakes can also be consumed faster. If a thicker shake is desired, simply use less liquid. Cut the liquid amount specified in the recipe in half. Add in more liquid as needed.

Read Food Labels

Read the nutrition facts label and look for the ingredients with the highest amount of calories. Look for "whole" fat da Avoid terms like "light", "skim", "low fat", or "diet" as these products are lower in calories. And we want all the calories we can get!

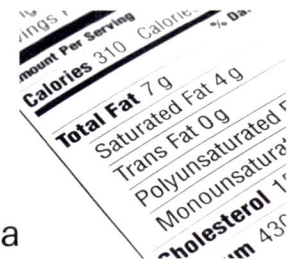

Ask about Favs

This may seem silly to even say, but ask about favorites. And don't just assume. Sometimes taste changes with age and when on certain medications. Maybe an old favorite doesn't taste good anymore. Ask them what their favorites are. Show them the recipes & pictures. Make note of their favorite flavors!

NON-DAIRY OPTIONS

All of the recipes in this cookbook feature dairy- it's a shake cookbook!

But what about those who can't do dairy?

For some people with lactose intolerance (inability to break down sugar in milk), they can take a **lactase enzyme** pill with their first sip of dairy. It is important that the lactase be taken with the dairy (not during scheduled pill times or after eating dairy). They can also purchase "lactose free" milk options.

Some people are unable to have dairy at all due to medical conditions or personal preference. Below are some non-dairy options for high calorie shakes.

INGREDIENTS FOR NON-DAIRY SHAKES

- **Plant based milks**
  - *Note:* sweetened versions, like vanilla, are often higher in calories than plain
  - Soy milk
  - Oat milk
  - Coconut milk
  - Rice milk
  - Almond milk (lower in calories)
- **Nut butters**
  - Peanut butter
  - Almond butter
  - Cashew butter
- **Plant based ice-creams**
  - Coconut milk based ice-cream
  - Soy milk based ice-cream
  - Other plant based ice-creams
- **Other plant-based foods**
  - Avocado
  - Olive oil
  - Coconut cream

Substitute high calorie non-dairy options for milk and ice-cream listed in these recipes. Note the calorie content will be different. Read the nutrition facts label for calorie content.

THE RECIPES

CLASSIC SERIES

PARTY SERIES

NUTTY SERIES

PIE SERIES

CLASSIC SERIES

VANILLA BEAN

INGREDIENTS

3/4 cup half and half

1 cup vanilla ice-cream

1 pkt. Carnation Instant Breakfast

Maraschino cherry, if desired

DIRECTIONS

- Add half-and-half, ice-cream, and Carnation Instant Breakfast into a blender.
- Blend ingredients until smooth.
- Pour into a glass and garnish with a cherry, if desired.

Calorie Boost: You can increase the calorie content by swapping out some of the half and half with whipping cream. Swapping out just 1/2 cup can add 260 extra calories! This much whipping cream will make the shake even creamier.

CALORIE COUNT:

625

14 grams of protein

CHOCOLATE SHAKE

INGREDIENTS

3/4 cup whole milk

1 cup chocolate ice-cream

1/4 cup chocolate syrup

Whipped cream, if desired

DIRECTIONS

- Add milk, ice-cream, and chocolate syrup into a blender.
- Blend ingredients until smooth.
- Pour into a glass and garnish with whiped cream, if desired.

Pro Tip: You can use different varieties of chocolate ice-cream to add a twist to this recipe- chocolate brownie, chocolate chocolate chip, chocolate swirl, etc.

CALORIE COUNT:

575

8 grams of protein

STRAWBERRY SHAKE

INGREDIENTS

3/4 cup whipping cream

1 cup vanilla ice-cream

1 cup strawberries

2 Tbsp granulated sugar

DIRECTIONS

- Remove the stems & tops from strawberries.
- Add whipping cream, ice-cream, strawberries, and sugar into a blender.
- Blend ingredients until smooth.

Pro Tip: If you don't have fresh strawberries, frozen work just as good! Or if you are out of berries, use strawberry ice-cream with the whipping cream.

CALORIE COUNT:

950

2 grams of protein

CARAMEL SHAKE

INGREDIENTS

3/4 cup whole milk

1 cup vanilla ice-cream

1/4 cup caramel sauce

Caramel drizzle, if desired

DIRECTIONS

- Add milk, ice-cream, and caramel sauce into a blender.
- Blend ingredients until smooth.
- Pour into a glass and drizzle with caramel, if desired.

Pro Tip: Home-made caramel sauce tastes especially delicious in this recipe. You can also add a dash of salt for a "salted caramel" variety!

CALORIE COUNT:

575

8 grams of protein

BANANA BLAST

INGREDIENTS

3/4 cup half and half

1 frozen banana

1 cup vanilla ice-cream

1/2 tsp vanilla extract

Whipped cream, if desired

DIRECTIONS

- Add half and half, banana, ice-cream, and vanilla extract into a blender.
- Blend ingredients until smooth.
- Pour into a glass and garnish with whipped cream, if desired.

Pro Tip: When bananas start "going bad", throw them sealed bag and put them into the freezer. Save them for milkshake recipes like this one!

CALORIE COUNT:

605

9 grams of protein

BUTTERSCOTCH SHAKE

INGREDIENTS

3/4 cup whole milk

1 cup vanilla ice-cream

1/4 cup butterscotch sauce

Whipped cream, if desired

DIRECTIONS

- Add milk, ice-cream, and butterscotch syrup into a blender.
- Blend ingredients until smooth.
- Pour into a glass and garnish with whipped cream, if desired.

Calorie Boost: Switch out 1/2 cup whole milk for sweetened condensed milk. This will male the shake super sweet, & it will add nearly 450 calories!!

CALORIE COUNT:

575

8 grams of protein

COFFEE SHAKE

INGREDIENTS

3/4 cup half and half

2 Tbsp instant coffee granules

1/4 cup dry whole milk powder

1 cup coffee ice-cream

Whipped cream, if desired

DIRECTIONS

- Add half and half, instant coffee, dry milk and ice-cream into a blender.
- Blend ingredients until smooth.
- Pour into a glass and garnish with whipped cream, if desired.

Pro Tip: Add 1/4 cup chocolate syrup to this recipe to turn this coffee shake into a cool refreshing mocha milkshake!

CALORIE COUNT:

650

15 grams of protein

MALTED MAGIC

INGREDIENTS

3/4 cup whole milk

1 cup vanilla ice-cream

1/4 cup malted milk powder

1/4 cup flavored syrup

Sprinkles, if desired

DIRECTIONS

- Add milk, ice-cream, malted milk powder, and flavored syrup into a blender.
- Blend ingredients until smooth.
- Pour into a glass and garnish with sprinkles, if desired.

Pro Tip: For this tasty treat you can make the malt ANY flavor. Just change the syrup flavor- chocolate, marshmallow, strawberry, caramel, fudge, etc.

CALORIE COUNT:

605

8 grams of protein

NUTTY SERIES

PEANUT BUTTER BLISS

INGREDIENTS

3/4 cup half and half

1 cup vanilla ice-cream

1/4 cup peanut butter

Whipped cream, if desired

DIRECTIONS

- Add half-and-half, ice-cream, and peanut butter into a blender.
- Blend ingredients until smooth.
- Pour into a glass and garnish with whipped cream, if desired.

Calorie Boost: Peanut butter is super high in heart healthy calories. Add in a little extra to boost the calories in this or any other milkshake.

CALORIE COUNT:

875

25 grams of protein

MAPLE NUT

INGREDIENTS

3/4 cup half and half

1 cup maple nut ice-cream

1 Tbsp maple syrup

2 Tbsp chopped walnuts

DIRECTIONS

- Chop walnuts well, unless you have a high powered blender.
- Add half-and-half, ice-cream, syrup, and walnuts into a blender.
- Blend ingredients until smooth.
- Pour into a glass and serve.

Pro Tip: If you are out of maple nut ice-cream, you can substitute this recipe with vanilla ice-cream. Add a little extra maple and/or walnuts as desired.

CALORIE COUNT:

650

9 grams of protein

PEANUT BUTTER CUP

INGREDIENTS

3/4 cup half and half

1 cup vanilla ice-cream

1/4 cup peanut butter

1/4 cup chocolate syrup

Chocolate chips, if desired

DIRECTIONS

- Add half-and-half, ice-cream, peanut butter, and choc. syrup into a blender.
- Blend ingredients until smooth.
- Pour into a glass and garnish chocolate chips, if desired.

Pro Tip: You can top this shake with chopped up peanut butter cups or even add 1-2 peanut butter cups to this milkshake for a fun twist.

CALORIE COUNT:

1060

25 grams of protein

ALMOND BUTTER

INGREDIENTS

3/4 cup half and half

1 cup vanilla ice-cream

1/4 cup almond butter

2 Tbsp honey

Whipped cream, if desired

DIRECTIONS

- Add half-and-half, ice-cream, almond butter, and honey into a blender.
- Blend ingredients until smooth.
- Pour into a glass and garnish with whipped cream, if desired.

Pro Tip: You can substitute any type of nut butter, like cashew or hazelnut or pistachio, with this recipe (or any recipe which calls for peanut butter).

CALORIE COUNT:

975

19 grams of protein

PEANUT BUTTER BANANA

INGREDIENTS

3/4 cup half and half

1 cup vanilla ice-cream

1/4 cup peanut butter

1 frozen banana

DIRECTIONS

- Add half-and-half, ice-cream, peanut butter, and banana into a blender.
- Blend ingredients until smooth.
- Pour into a glass and serve.

Pro Tip : You can add 1/4 cup chocolate syrup to this recipe to make it "The King of Calories", the shake with the most calories in this cookbook!

CALORIE COUNT:

1000

26 grams of protein

HAZELNUT CHOCOLATE

INGREDIENTS

3/4 cup half and half

1 cup vanilla ice-cream

1/4 cup hazelnut chocolate spread

Whipped cream, if desired

DIRECTIONS

- Add half-and-half, ice-cream, and hazelnut chocolate spread into a blender.
- Blend ingredients until smooth.
- Pour into a glass and garnish with whipped cream, if desired.

Pro Tip: You can swap out the hazelnut chocolate spread with any other dessert spread, such as biscoff or cookie butter spreads, for a fun twist.

CALORIE COUNT:

890

11 grams of protein

PEANUT BUTTER & JELLY

INGREDIENTS

3/4 cup whole milk

1/4 cup peanut butter

1/8 cup jelly or jam

1 cup vanilla ice-cream

Extra jelly or jam, if desired

DIRECTIONS

- Add milk, peanut butter, jam or jelly, and ice-cream into a blender.
- Blend ingredients until smooth.
- Pour into a glass and garnish with a dollop of jelly or jam, if desired.

Pro Tip: Use any flavor of jelly or jam desired- strawberry, grape, apricot, blackberry, etc. Short on jam? Use 1/4 cup fruit and 2 Tbsp sugar instead.

CALORIE COUNT:

800

26 grams of protein

PARTY SERIES

AVOCADO SHAKE

INGREDIENTS

3/4 cup whole milk

3/4 cup mashed avocado

1 cup vanilla ice-cream

2 Tbsp granulated sugar

whip cream, if desired

DIRECTIONS

- Add milk, avocado, ice-cream, and sugar into a blender.
- Blend ingredients until smooth.
- Pour into a glass and garnish with whip cream, if desired.

Pro Tip: Avocados give a surprising creamy and delicious taste to shakes. Try adding 1/4 cup chocolate syrup for another fun variation!

CALORIE COUNT:

750

11 grams of protein

PINA COLADA

INGREDIENTS

3/4 cup half and half

1/4 cup canned coconut cream

1 cup vanilla ice-cream

1/2 cup crushed pineapple

Whipped cream, if desired

DIRECTIONS

- Add half & half, coconut cream, ice-cream, and pineapple into a blender.
- Blend ingredients until smooth.
- Pour into a glass and garnish with whipped cream, if desired.

Pro Tip: If you really want to make this a fancy drink, you can add whipped cream, shredded coconut, and cherry on top! Or even a drink umbrella!

CALORIE COUNT:

690

8 grams of protein

CHOCOLATE MINT

INGREDIENTS

1/2 cup whole milk

1/2 cup whipping cream

1 1/2 cups mint chocolate chip ice-cream

Shaved chocolate, if desired

DIRECTIONS

- Add milk, whipping cream, and ice-cream into a blender.
- Blend ingredients until smooth.
- Pour into a glass and garnish with shaved chocolate, if desired.

Pro Tip: For an extra creamy and "fluffy" shake, blend the whipping cream first until it starts to thicken. Then add milk and ice-cream. Blend until smooth!

CALORIE COUNT:

850

7 grams of protein

CHOCOLATE BANANA

INGREDIENTS

3/4 cup whole milk

1 cup chocolate ice-cream

1 frozen banana

1/4 cup chocolate syrup

DIRECTIONS

- Add milk, ice-cream, banana, and chocolate syrup into a blender.
- Blend ingredients until smooth.
- Pour into a glass and serve.

Calorie Boost: If you would like to add even more calories to this milkshake, swap out the whole milk for half and half or whipping cream.

CALORIE COUNT:

680

9 grams of protein

COCONUT CREAM

INGREDIENTS

3/4 cup whole milk

1/3 cup canned coconut cream

1 cup vanilla ice-cream

1/2 tsp. coconut extract

shredded coconut, if desired

DIRECTIONS

- Add milk, coconut cream, ice-cream, and coconut extract into a blender.
- Blend ingredients until smooth.
- Pour into a glass and garnish with shredded coconut, if desired.

Pro Tip: Want a COLDER shake- scoop coconut cream into an ice cube tray. Freeze until sold and make shake with coconut cream ice cubes!

CALORIE COUNT:

500

9 grams of protein

CINNAMON ROLL

INGREDIENTS

1/2 cup whole milk

1/4 marshmallow fluff

1/8 cup caramel syrup

1 cup vanilla ice-cream

1/2 tsp. cinnamon

Whipped cream, if desired

DIRECTIONS

- Add milk, marshmallow, caramel, ice-cream, and cinnamon into a blender.
- Blend ingredients until smooth.
- Pour into a glass and garnish with whipped cream & cinnamon , if desired.

Pro Tip: You can add chunks of real cinnamon roll into this milk shake to give it more flavor and texture! Add more or less cinnamon based on preference.

CALORIE COUNT:

530

5 grams of protein

40

PIE SERIES

APPLE PIE

INGREDIENTS

1/4 cup whole milk

1 cup canned apple pie filling

1 cup vanilla ice-cream

4 vanilla wafer cookies

Whipped cream, if desired

DIRECTIONS

- Add milk, pie filling, ice-cream, and cookies into a blender.
- Blend ingredients until smooth.
- Pour into a glass and garnish with whipped cream and crumbled cookies, if desired.

Pro Tip: Do you like making home-made pie? You can use your own pie recipe (baked apple, cinnamon, nutmeg, sugar) as the filling for this recipe!

CALORIE COUNT:

625

4 grams of protein

CHERRY PIE

INGREDIENTS

1/4 cup whole milk

1 cup canned cherry pie filling

1 cup vanilla ice-cream

4 vanilla wafer cookies

Extra pie filling, if desired

DIRECTIONS

- Add milk, pie filling, ice-cream, and cookies into a blender.
- Blend ingredients until smooth.
- Pour into a glass and garnish with extra pie filling, if desired.

Calorie Boost: Swap out the whole milk with half and half or whipping cream to add an extra boost of calories and creaminess.

CALORIE COUNT:

625

4 grams of protein

PUMPKIN PIE

INGREDIENTS

1/4 cup sweetened condensed milk

1/4 cup whole milk

1 cup vanilla ice-cream

1/2 cup canned pumpkin

1/2 tsp. pumpkin pie spice

Whipped cream, if desired

DIRECTIONS

- Add sweetened condensed milk, whole milk, ice-cream, pumpkin, and pumpkin pie spice into a blender.
- Blend ingredients until smooth.
- Pour into a glass and garnish with whipped cream, if desired.

Pro Tip: If you want a milder taste only use 1/4 tsp. pumpkin pie spice.

CALORIE COUNT:

710

11 grams of protein

BLUEBERRY PIE

INGREDIENTS

1/4 cup whole milk

1 cup canned blueberry pie filling

1 cup vanilla ice-cream

4 vanilla wafer cookies

DIRECTIONS

- Add milk, pie filling, ice-cream, and cookies into a blender.
- Blend ingredients until smooth.
- Pour into a glass and serve.

Pro Tip: You can also use fresh or frozen blueberries with this recipe. Substitute the canned mix for fresh or frozen berries. Add in 2 Tbsp. sugar.

CALORIE COUNT:

625

4 grams of protein

Weight Gaining
SECRETS

BONUS

Enjoy this BONUS copy of my Weight Gain SECRETS book.

This book is intended to compliment your High Calorie SHAKES Cookbook. It provides the tips and tools you need to successfully gain weight (and stop unintended weight loss!).

Weight Gain SECRETS will cover:
- The 9 Secrets to Gaining Weight
- High Calorie Diet Planning Tools
- BONUS High Calorie Drink Recipes

Enjoy.

THE 9 SECRETS TO GAINING WEIGHT

TIP 1 Eat more calories, not just more food.

It's a common misconception that to gain weight you just need to eat more food. But that is easier said than done, right?

The reality is that you need more calories. Calories are the key to packing on the pounds. Different types of foods have different amounts of calories. If you just eat "more food" blindly, this does not mean you will gain weight.

In fact, some types of food (ex. food high in fiber) , can actually make you feel full fast. Then it's even harder to eat more food.

So remember, eat more calories. Not just more food. Later in this book we will have everything you need to know about high calorie foods!

TIP 2 Fat has more calories than other nutrients.

Calories come from macronutrients: protein, carbohydrates, and fat. But when it comes to calories, these three types of nutrients are not created equal.

Here is the breakdown of calories per gram with each nutrient:

- **Protein:** 4 calories per gram
- **Carbohydrates:** 4 calories per gram
- **Fat:** 9 calories per gram

Did you catch that? Fat has more than twice the calories per volume than protein and carbohydrates. Remember tip #1: Eat more calories, not just more food? Foods high in fat contain the most calories.

This is most helpful for those who have poor appetite and can't eat large volumes of food. Eating high fat foods contain more calories per volume. Fat is found in oils, full fat dairy, meat, and even avocados.

Keep in mind, all nutrients are important for health. Some individuals may have medical conditions or nutrition needs that require less fat. Talk to your healthcare team about your individual needs.

Also note that there are different types of fat- monounsaturated and polyunsaturated fats are heart healthy choices. Saturated fats should be limited and trans fats should be avoided all together.

TIP 3 Know how to read a nutrition facts label.

At this point you may be wondering- how do I find foods high in calories and/or fat? To do this, you need to know how to read a nutrition facts label.

The truth is that foods are not made of a single nutrient. They are made of different combinations of macronutrients. And prepared foods have a variety of food ingredients to mix things up even more.

When reading a nutrition facts label there is ONE thing you need to look at first: Serving Size! A serving size may be a quarter cup, a full cup, or even the entire container.

After you figure out the serving size, look at how many calories are provided in that amount. You can also find out how much fat and which type it is (again, it's often a combination).

If you eat more than one serving size, do the math to figure out how much you are actually consuming.

Knowing how to read a nutrition facts label puts you ahead of the game. Ensuring you are buying and eating the foods you need with the most calories!

TIP 4 Eat smaller volumes of food more often.

Remember tip #1 where we said eat more calories, not just more food? One of the best ways to eat more calories is to eat "calorie dense" foods more often throughout the day.

"Calorie dense" foods are those foods highest in calories, typically high in fat like we discussed in tip #2. Most people will hit a point where they are too full and just can't eat anymore darn food.

Being strategic in eating calorie dense foods in smaller volumes, but more often is a secret to get in more calories throughout the day.

To do this you can think of eating 6 small meals per day or 3 meals plus 3 snacks. However you like to think of eating.

TIP 5 High calorie drinks are your secret weapon.

My favorite weight gaining secret is to consume high calorie drinks. I've found that even when someone cannot eat another bite of food, they can still sip on a high calorie beverage.

High calorie drinks are your secret weapon to gaining weight! You can drink them in between meals to add even more calories to your day. And there is such a variety to choose from!

Some of the top high calorie drinks include:
- High calorie shakes
- High calorie smoothies
- Fortified milk
- High calorie cocoa
- Pre-made nutrition shakes

You can find recipes for the first four options later in this book. Pre-made nutrition shakes are a good option to keep on hand in case you aren't up to making a drink from scratch.

Read the labels on your beverages to find the highest calorie options. When making smoothies or milkshakes use frozen milk or juice instead of plain ice/water.

TIP 6 Gain weight slowly and exercise.

I know it's tempting to want to gain a lot of weight really fast. But it's more important to gain weight the right way. And in an ideal world you will be gaining muscle along with your body weight.

You should aim at gaining weight at a rate of no more than one pound per week. Gaining too much weight too fast can negatively impact your health and/or lead to increases in fat mass over muscle mass.

This is where exercise and gaining weight slowly come together. You need to be doing regular resistance training exercise to protect and grow your muscles. Muscle mass is so important for good health.

Here's the hard part. When you exercise you burn calories. This is a bummer because you are trying so hard to eat enough calories. But you absolutely need to exercise.

You do not want to gain a large amount of fat mass and have no muscle to support your added weight. The goal with weight gain should always be improved health and increased function (read= muscle).

Be sure to talk to your healthcare team prior to starting any type of exercise program.

TIP 7 Get enough protein to protect your muscles.

Let's talk more about muscle, since this is so important. In addition to getting resistance training for muscle health, you need to be getting enough protein. Protein is found in meat, dairy, and a variety of foods.

Protein + Resistance Training = Happy Muscles

How much protein do you need? Most people need 0.8 grams of protein per kilogram body weight (there are 2.2 kilograms in a pound). Here's the math:

(weight in pounds/2.2) x 0.8 = grams of protein you need daily

Some people may need more, perhaps 1.0-1.2+ grams/kg body weight. Always be sure to check with your healthcare team before making changes to your diet. Some people do need less protein.

TIP 8 Address underlying medical issues.

If you are losing weight or struggling to gain weight, your first step should always be to consult with your healthcare team. I hope you've already done this, but if not make an appointment now.

There can be an underlying medical issue that is causing you to lose weight despite eating more calories and trying to gain weight.

The last thing you want to do is mask a serious medical issue by eating a high calorie diet without addressing the underlying issue. Sometimes just addressing that underlying issue can make it way easier to get back to gaining weight!

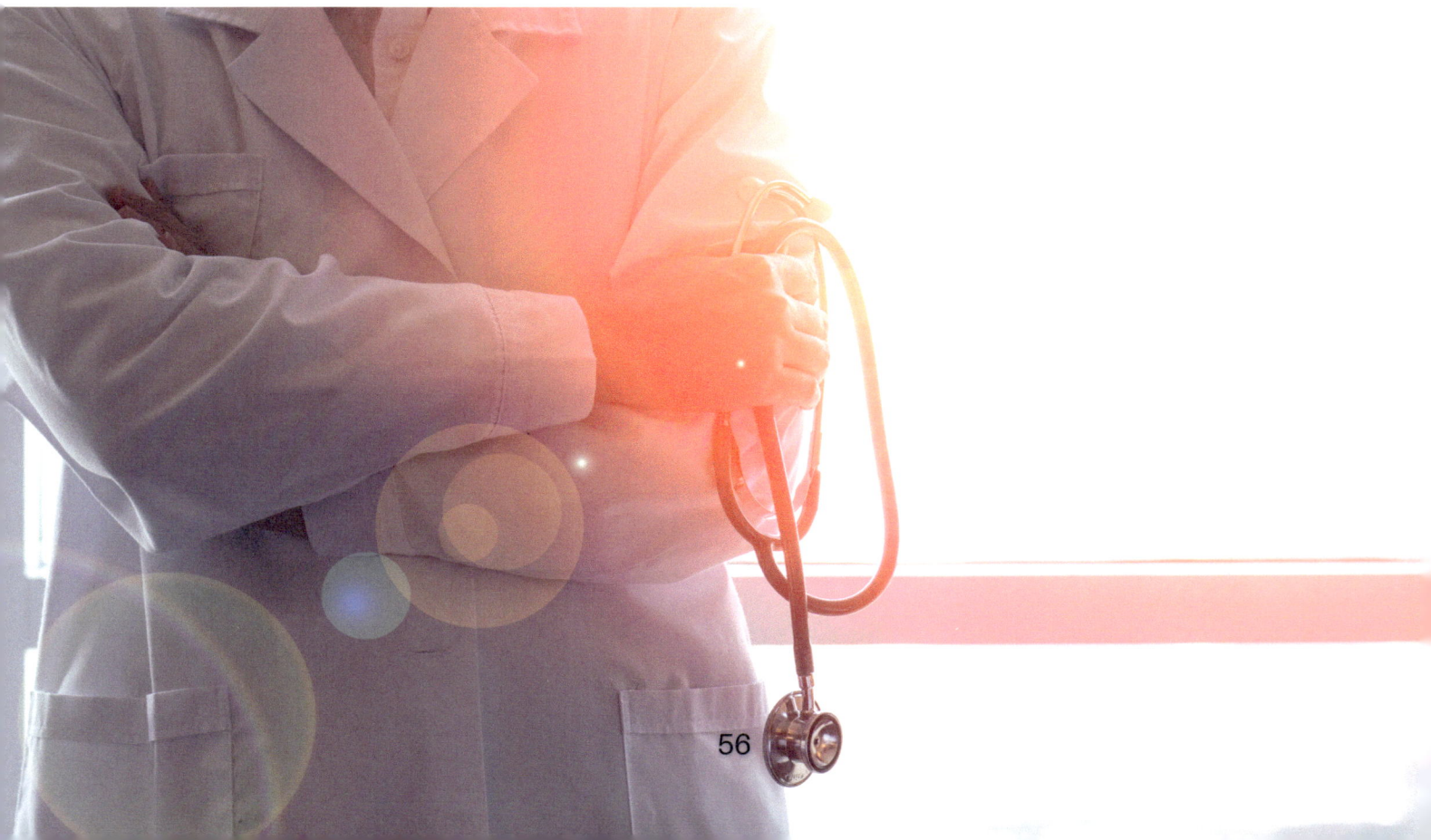

TIP 9 Plan ahead for continued success.

The best way to success is to plan ahead. In the following pages you will learn all about which foods are highest in calories.

Take advantage of the high calorie grocery list and high calorie meal planning worksheets. Figure out your favorite high calorie foods and make a plan.

Those who plan ahead have the most success. They make their dreams a reality and don't just wonder why they aren't gaining weight. Make a plan and stick to it. You can do this!

THE 9 SECRETS TO GAINING WEIGHT

TIP SUMMARY

1) Eat more calories, not just more food.
2) Fat has more calories than other nutrients.
3) Know how to read a nutrition facts label.
4) Eat smaller volumes of food more often.
5) High calorie drinks are your secret weapon.
6) Gain weight slowly and exercise.
7) Get enough protein to protect your muscles.
8) Address underlying medical issues.
9) Plan ahead for continued success.

Now that you know the top 9 secrets to gaining weight, let's review some tools you can use to plan ahead for continued success. We'll also look some high calorie drink recipes and calorie counts!

HIGH CALORIE DIET PLANNING TOOLS

HIGH CALORIE FOODS
to stop weight loss & promote weight gain.

DAIRY
Whole milk
Half and half
Buttermilk
Chocolate milk
Powdered milk
Greek yogurt
Cottage cheese
Cream cheese
Ice cream

Schedule

SWEETS
Syrup
Honey
Sugar
Agave
Jam
Hazelnut spread
Flavored syrups

FATS
Avocado
Olive oil
Butter
Nuts
Nut butters
Soft spread margarine
Egg yolks

OTHERS
Dried fruit
Coconut
Pre made nutrition shakes
Protein Powder
Flax seed

High Calorie Diet Tips:
Serve high calorie foods in meals and snacks.
Add high calorie foods to the foods someone is already eating (ex. mix into oatmeal, soups, smoothies, sauces, etc.).
Be sure to read labels for nutrition information (calories may vary by recipe/brand).

High Calorie Foods
GROCERY LIST

PRODUCE
Avocado
Coconut

FREEZER
Ice cream
Frozen custard

PANTRY
Nuts
Nut butters
Jam
Hazelnut spread
Dried fruit
Olive oil
Protein powder
Pre made nutrition shakes
Flax seed

DAIRY
Whole milk
Half and half
Buttermilk
Chocolate milk
Cottage Cheese
Greek Yogurt
Cream Cheese
Soft spread margarine
Butter
Eggs

BAKING
Powdered milk
Maple syrup
Flavored syrups
Honey
Agave

I ALSO NEED:

HIGH CALORIE MEAL PLANNER
This tool has high calorie foods and add-ins. Add in your normal meals to complete a meal plan unique to your needs.

BREAK-FAST	Whole milk	Whole milk	Whole milk	Whole milk	Whole milk
SNACK	High calorie shake	Peanut butter Toast	High calorie shake	Peanut butter Apples	High calorie shake
LUNCH	Butter or olive oil	Butter or olive oil	Butter or olive oil	Butter or olive oil	Butter or olive oil
SNACK	Cottage cheese Canned fruit	Hard boiled egg Tomatoes	Whole milk Cookies	Avocado Toast	Greek yogurt Blueberries
DINNER	Butter or olive oil	Butter or olive oil	Butter or olive oil	Butter or olive oil	Butter or olive oil
SNACK	High calorie shake	High calorie shake	High calorie shake	High calorie shake	High calorie shake
SNACK					

HIGH CALORIE FOODS

to stop weight loss & promote weight gain.

DAIRY

Whole milk
Half and half
Buttermilk
Chocolate milk
Powdered milk
Greek yogurt
Cottage cheese
Cream cheese
Ice-cream

FATS

Avocado
Olive oil
Butter
Nuts
Nut butters
Soft spread margarine
Egg yolks

SWEETS

Syrup
Honey
Sugar
Agave
Jam
Hazelnut spread
Flavored syrups

OTHERS

Dried fruit
Coconut
Pre-made nutrition shakes
Protein Powder
Flax seed

High Calorie Diet Tips:

-Serve high calorie foods in meals and snacks.
-Add high calorie foods to the foods someone is already eating (ex. mix into oatmeal, soups, smoothies, sauces, etc.).
-Be sure to read labels for nutrition information (calories may vary by recipe/brand).

High Calorie Foods
GROCERY LIST

PRODUCE
Avocado
Coconut

FREEZER
Ice-cream
Frozen custard

PANTRY
Nuts
Nut butters
Jam
Hazelnut spread
Dried fruit
Olive oil
Protein powder
Pre-made nutrition shakes
Flax seed

DAIRY
Whole milk
Half and half
Buttermilk
Chocolate milk
Cottage Cheese
Greek Yogurt
Cream Cheese
Soft spread margarine
Butter
Eggs

BAKING
Powdered milk
Maple syrup
Flavored syrups
Honey
Agave

I ALSO NEED:

HIGH CALORIE MEAL PLANNER

This tool has high calorie foods and add-ins. Add in your normal meals to complete a meal plan unique to your needs.

BREAK-FAST	Whole milk	Whole milk	Whole milk	Whole milk	Whole milk
SNACK	High calorie shake	Peanut butter Toast	Peanut butter Apples		High calorie shake
LUNCH	Butter or olive oil	Butter or olive oil	Butter or olive oil	Butter or olive oil	Butter or olive oil
SNACK	Cottage cheese Canned fruit	Hard-boiled egg Tomatoes	Whole milk Cookies	Avocado Toast	Greek yogurt Blueberries
DINNER	Butter or olive oil	Butter or olive oil	Butter or olive oil	Butter or olive oil	Butter or olive oil
SNACK	High calorie shake	High calorie shake	High calorie shake	High calorie shake	High calorie shake

Visit **highcaloriemealplanner.com** to download a printable version of the meal planner.

62

HIGH CALORIE MEAL PLANNER

Add in your normal meals and high calorie foods to complete a meal plan unique to your needs.

BREAK -FAST					
SNACK					
LUNCH					
SNACK					
DINNER					
SNACK					

Visit **highcaloriemealplanner.com** to download a printable version of the meal planner.

63

HIGH CALORIE DRINKS RECIPES

Best High Calorie Drinks

HIGH CAL SHAKES **HIGH CAL SMOOTHIES** **FORTIFIED MILK** **HIGH CAL COCOA** **NUTRITION SHAKES**

You've got the High Calorie Shakes! Check out our recipes for:
- High Calorie Smoothies
- Fortified Milk
- High Calorie Cocoa

For pre-made nutrition shakes, any brand will do. Make sure you read the nutrition facts label. Protein drinks aren't typically as high in calories as nutrition shakes. "Plus" varieties have the most calories.

High Calorie Strawberry Banana Smoothie

650 calories

- 2 c. strawberries
- 1 medium banana
- ½ c. Greek yogurt
- 1 Tbsp honey
- 1 Tbsp chia seed
- 1 c. whole milk

Blend until smooth!

High Calorie Pineapple Coconut Smoothie

525 calories

- 1 c. pineapple juice
- 1 med. banana
- 1/2 c. coconut cream
- 1 c. pineapple

Blend until smooth!

High Calorie Green Smoothie

515 calories

- 1 c. apple juice
- 1 c. spinach
- 1 med. apple
- 1 med. avocado
- 1 cup pineapple

Blend until smooth!

High Calorie Peanut Butter Oatmeal Smoothie

730 calories

- ¼ cup oats
- 1 med. banana
- 1/2 c. half & half
- ¼ c. peanut butter
- 1 Tbsp honey

Blend until smooth!

FORTIFIED MILK

Fortified milk, also known as "double strength" milk is an easy and simple high calorie drink option. You simply combine liquid milk with powdered milk. You are getting a condensed version of milk, packed with calories. Drink plain or add to other foods/beverages you eat.

Fortified Milk Recipe

Ingredients:
- 1 cup whole milk
- 1/4 cup dry milk (whole)

Directions:
Combine ingredients and stir until smooth. You can also mix the ingredients in a shaker.

Calorie Count: 307

HIGH CALORIE COCOA

Who doesn't love some classic cocoa? Served hot or chilled. With whipped cream or marshmallows. This is a tasty high calorie options that can be served in the morning or evening (or anytime!).

Tip: Add flavored syrups to change the flavor (ex. raspberry, mint, etc.).

High Calorie Cocoa Recipe

Ingredients:
- 1 cup half and half
- 2 Tbsp chocolate syrup
- 2 Tbsp dry milk (whole)

Directions:

If desired, heat half and half first. Mix in chocolate syrup and dry milk until smooth. Pour into a mug and top with whipped cream or marshmallows and chocolate if desired.

Calorie Count: 500

CALORIE COUNTS

Calories in Milk (serving size: 1 cup)

- Milk, whole- 149
- Milk, whole chocolate- 204
- Milk, whole strawberry- 201
- Milk, 2%- 122
- Milk, 1%- 102
- Milk, Skim- 83
- Half and Half- 320
- Buttermilk- 152
- Kefir, sweetened- 150
- Goats milk, whole- 168

Calories in Juice (serving size: 1 cup)

- Grape Juice- 152
- Cranberry Juice- 115
- Orange Juice- 110
- Apple Juice- 110
- Tomato Juice- 50

Calories in Plant-Based Milks (serving size: 1 cup)

- Soy milk- 131
- Oat milk- 120
- Rice milk- 112
- Coconut milk- 75
- Almond milk, unsweetened- 92
- Almond milk, sweetened- 37

CALORIE COUNTS

In conclusion, I wanted to leave you with my top 15 high calorie vegan foods list! These are nutritious food options for anyone. But note this page includes only plant-based high calorie food options.

Calories in Plant-Based Foods

- **Walnuts, chopped** (¼ cup) **191 calories**
- **Peanut butter** (2 Tbsp) **188 calories**
- **Avocado** (1 cup) **240 calories**
- **Dried plums** (½ cup) **209 calories**
- **Coconut cream** (½ cup) **240 calories**
- **Vegetable oil spread** (2 Tbsp) **164 calories**
- **Soy milk** (1 cup) **131 calories**
- **Olive oil** (2 Tbsp) **238 calories**
- **Rice, white, cooked** (1 cup) **206 calories**
- **Quinoa, cooked** (1 cup) **222 calories**
- **Potato, baked** (1 medium) **163 calories**
- **Flax seed, ground** (¼ cup) **148 calories**
- **Honey** (¼ cup) **256 calories**
- **Dates** (½ cup) **207 calories**
- **Pumpkin Seeds** (½ cup) **142 calories**

NOTES

NOTES

NOTES

NOTES

NOTES

NOTES

NOTES

NOTES

Thank you!

Learn more about geriatric nutrition at
www.thegeriatricdietitian.com

The Geriatric Dietitian
FOOD + LOVE

Learn more about high calorie foods at
www.highcalorierecipes.com

High Calorie
RECIPES

Visit **highcaloriemealplanner.com** to download a printable version of the meal planner provided in this book.

Made in United States
Troutdale, OR
09/21/2024

23021320R00050